Turbulence and random processes in fluid mechanics,
second edition

Turbulence and random processes in fluid mechanics
Second edition

M. T. LANDAHL
Massachusetts Institute of Technology

E. MOLLO-CHRISTENSEN
Formerly of Massachusetts Institute of Technology

CAMBRIDGE
UNIVERSITY PRESS

Published by the Press Syndicate of the University of Cambridge
The Pitt Building, Trumpington Street, Cambridge CB2 1RP
40 West 20th Street, New York, NY 10011-4211, USA
10 Stamford Road, Oakleigh, Victoria 3166, Australia

First published 1986
Second edition 1992
Printed in the United States of America

Library of Congress Cataloging-in-Publication Data
Landahl, Mårten.
Turbulence and random processes in fluid mechanics / M. T. Landahl,
E. Mollo-Christensen. – 2nd ed.
p.   cm.
Includes bibliographical references and index.
ISBN 0-521-41992-1 (hc). – ISBN 0-521-42213-2 (pb)
1. Fluid mechanics.   2. Turbulence.   3. Stochastic processes.
I. Mollo-Christensen, E.   II. Title.
QA901.L278   1992
532.0527 – dc20   91-44918

A catalog record for this book is available from the British Library.

ISBN 0-521-41992-1 hardback
ISBN 0-521-42213-2 paperback

Contents

Preface to the second edition

In addition to correcting misprints and errors in the text, the equations, and the figures of the first edition, we also have further clarified points that have proved difficult for students. We also have benefited from reviews of the book and made other additions and changes as needed.

We added in Chapter 8 a short description of a simplified model for the temporal and spatial evolution of three-dimensional disturbances in a strong mean shear, which we thought might give some theoretical framework for the study of bursting in the near-wall region of a turbulent boundary layer. We also have added a short chapter (Chapter 12) on numerical modeling of turbulence, the lack of which many reviewers pointed to as a shortcoming of the first edition.

A few reviewers have questioned the need to include stability and wave motions in an introductory book on turbulence. In our view, research on hydrodynamic instability has contributed significantly to our understanding of how turbulence is created and maintained. The work in the new field of nonlinear dynamical systems and their chaotic behavior has added further insights showing, for example, that nonlinear waves may show chaotic behavior.

Additions notwithstanding, we have tried strenuously to retain the compactness of the book. It is intended to be a graduate-level introduction and overview of the subject suitable for a one-term course.

Preface to the first edition

Fluid flow turbulence is a phenomenon of great importance in many fields of engineering and science. It presents some of the most difficult problems both in the fundamental understanding of its physics and in applications, many of which are still unresolved. Turbulence and related areas have therefore continued to be subjects of intensive research over a period that has lasted for more than a century, and the interest in this field shows no signs of abatement.

In recognition of the need for helping graduate students prepare for their own research in this and related areas of fluid dynamics, a course with the cover title was started by one of us (E. M.-C.) some 20 years ago. Our joint efforts in producing a set of notes for this course has resulted in the present book. The course and its subject matter has evolved over this time period of teaching a mixed group of students from all fields of engineering and from many areas of science, including astrophysics, physics, chemistry, applied mathematics, meteorology, oceanography, and occasionally biology and physiology. With students of such widely different backgrounds we could not assume much commonality in preparation beyond the basics. Hence we found it necessary to start each topic at a fundamental level, and very few concepts could be borrowed from common professional experiences. Many of the students in the course were looking for a thesis topic or needed more insight into turbulence in support of their ongoing research. Discussions with students have resulted in the start of successful research subjects in many instances.

The main aim of the book is to give the students the background enabling them to follow the literature and understand current research results. The book stresses fundamental concepts and basic methods and approaches, although attempting to introduce some recent ideas that we think will prove important in future work on turbulence and related fields. The flavor of a course based on this book will be strongly dependent on the instructor and on the emphasis and the examples of research results chosen for presentation, since the book in itself is not a complete course. Reading of the literature and monographs are also needed. We have to some extent departed from the conventional treatment of the subject, which is well covered in earlier monographs and textbooks such as those by Batchelor (1953), Townsend (1956, 1976), Hinze (1975), Monin and Yaglom (1972), and Tennekes and Lumley (1972). Our treatment stresses the dynamics of processes that participate in creating and maintaining turbulent flows. Among such topics are stability, wave motion, and coherent structures. Although not covered in detail, the "standard" equilibrium turbulent flows are also described (i.e., channel flow, boundary layers, and convection between two parallel plates).

We acknowledge the many contributions our students have made to the material in the book. Special thanks go to Professor Joseph Haritonidis for his contributions in teaching portions of this course and his many suggestions for improving the text, to Professor Sheila Widnall for her invaluable technical advice and help, to Dr. Arne Johansson for his careful reading of the final manuscript, to Christine Collins and Linda Notzelman for their assistance with the preparation of the manuscript, and to Catherine Lagercrantz for her assistance in completing the book by finishing references and drawings and putting it all together.

Cambridge, Mass.   M. L.   E. M.-C.

1

Introduction with historical notes

Most flows encountered in nature and in engineering practice are turbulent. It is therefore important to understand the fundamental mechanisms at work in such flows. Turbulent flows are unsteady and contain fluctuations that are random in space and time. An important characteristic is the richness of scales of eddy motion present in such flows: In a fully developed turbulent flow all scales appear to be fully occupied or saturated in a sense, from the largest ones that can fit within the size of the flow region down to the smallest scale allowed by dissipative processes. Turbulent flows are also highly vortical, a consequence of vortex stretching and tilting by larger random vorticity fields.

The reason turbulence is so prevalent in fluids of low viscosity is that steady laminar flows tend to become unstable at high Reynolds or Rayleigh numbers and therefore cannot be maintained indefinitely as steady laminar flows. Instability to small disturbances is an initial step in the process whereby a laminar flow goes through transition to turbulence. In investigations of instability of flows that are homogeneous in one or more spatial dimensions, one usually formulates a linear problem of the evolution of an infinite train of small-amplitude waves so as to find whether such waves will grow or decay with time. More general disturbances may be analyzed by Fourier superposition.

A complete description of the transition process requires one to consider the development of disturbances of finite amplitudes. This is generally a difficult theoretical task since it leads to nonlinear problems. A few simplified model problems, giving some insight into the nature of the transition process, are tractable, however, such as the evolution of a finite-amplitude wave train in a parallel shear flow, finite-amplitude density interface waves, weak nonlinear interaction of several wave trains, and the influence of distortion by large-scale motion on smaller-scale wave trains. The treatment of non-wave-like amplitude disturbances lacking spatial and temporal periodicity is more difficult.

Because of the random nature of fully developed turbulent flow fields, statistical methods are usually employed for their description. However, in the statistical averages much of the information that may be relevant to the understanding of the turbulent mechanisms may be lost, especially phase relationships. This may not seem too serious for flows in which the motion appears to be completely disorganized, such as in nearly isotropic or homogeneous turbulent flows. However, in order to understand highly intermittent turbulence production mechanisms for which intricate phase relationships are likely to play an essential role, standard averaging techniques are insufficient, and conditional sampling methods must be resorted to in order to bring out the important patterns in the fluctuating flow field. In recent years such sampling techniques have been employed in experimental investigations for both wall-bounded and free-shear flows, and these have given much new information on the structure of turbulence.

Turbulence and the closely related field of transition from a laminar to a turbulent flow have been the subject of serious scientific inquiry for about a century. It is a testimony to the difficulty of the subject that it is still far from exhausted and in fact rather incompletely understood in many of its fundamentals.

The first serious studies of flow instability and turbulence date from the late nineteenth century with the pioneering contributions by Osborne Reynolds and Lord Rayleigh. Reynolds's (1883) famous investigations of pipe flow clearly established that there are two fundamentally different modes of flow, laminar and turbulent (which he called "sinuous"). He demonstrated that an important nondimensional parameter that determines what kind of flow can be sustained is

$$\mathrm{Re} = \frac{Ud}{\nu},$$

where U is the (bulk) velocity, d the pipe diameter, and ν the kinematic viscosity, a parameter that later became known as the Reynolds number. Reynolds found that turbulent flow could be sustained only for Re greater than about 2300, a value known as the *critical Reynolds number*. Another of his important discoveries was that turbulence tended to set in intermittently in localized regions, "flashes", a property later found to be quite common in wall-bounded turbulence. Reynolds also made an important contribution to turbulence theory with his derivation of the equations of motion for the turbulent mean flow, which in particular led to the introduction of the fictitious so-called Reynolds stresses (1894) to represent the mean momentum fluxes due to turbulence.

At about the same time as Reynolds carried out his experiments with pipe flows, Lord Rayleigh started his theoretical investigations of the stability of a parallel flow of an inviscid fluid. His basic approach, which has been followed in most later investigations, was to determine whether a small disturbance in the form of an infinite wave train of uniform amplitude would grow or decay with time. His first publication on this subject dates from 1878. A very important result was his (1880) proof that a necessary condition for a parallel flow to be unstable is that its velocity profile have an inflection point.

The idea that a turbulent flow may be treated as a laminar flow with different fluid properties, the basic idea behind Reynolds's (1894) introduction of the fictitious turbulent stresses, was in fact put forward already by Boussinesq (1877). He assumed that the turbulent shear stress is proportional to the velocity gradient, just as viscous shear stresses in a laminar flow, but with a factor of proportionality different (generally much higher) than in the corresponding laminar case. The use of a turbulent viscosity ("eddy" viscosity) has been common in approximate theories for the mean flow. A constant value of turbulent viscosity has been found to lead to a fairly good representation of the mean velocity distribution in free turbulent flows (wakes, etc.), whereas for wall-bounded flows (pipe flows, channel flows, boundary layers) it is necessary to employ more refined hypotheses regarding how the turbulent viscosity varies with the distance from the wall. It was not until the works by Prandtl and von Karman in the period 1920–30 that methods of calculations were developed with which one was able to obtain approximations, based on a small number of empirical parameters, to the mean velocity field in reasonably good agreement with experiments. Particularly successful was Prandtl's (1925) introduction of the concept of a "mixing length" [also employed in an earlier work by Taylor (1915)], which measures the average distance a fluid element would stray from the mean streamline. Prandtl set down an approximate expression for the mean momentum transfer by turbulent mixing in terms of the mixing length. By assuming the mixing length to be proportional to the distance from the wall, one may then determine the mean velocity distribution near the wall with the aid of only one empirical constant. After Prandtl many different and more complicated hypotheses regarding the momentum transfer by the turbulent fluctuations have been put forward, but today there is not yet any method available for the practical calculation of the turbulent mean field based directly on the Navier–Stokes equations without the use of empirical data. The fundamental difficulty is that the fluctuating field depends in a nonlinear fashion on the mean velocity distribution, which in turn is governed by the Reynolds equations containing the mean momentum transfer by the fluctuating velocities.

The early approach to turbulence treated homogeneous and isotropic flows, building on ideas borrowed from the kinetic theory of gases. In such turbulent fields the velocity and pressure fluctuations are statistically independent of position and show no directional dependence. This simplifies their statistical description significantly. An idealized model of this kind represents fairly well the behavior of the turbulence in a large container, in which the fluid has been subjected to random stirring and then is left to equilibrate. More importantly, the smallest eddies in all turbulent flows are found to be approximately isotropic.

Many of the fundamental ideas in the statistical theory of turbulence were formulated early by G. I. Taylor (1923, 1935). He introduced the concept of velocity covariances (correlations),

$$R_{ij} = \langle u_i(x_k) u_j(x_k + \xi_k) \rangle,$$

between one velocity component u_i (Cartesian tensor notation being used) at the point x_k with another component u_j (where i and j may be equal or different) at the point $x_k + \xi_k$. The angular brackets $\langle\ \rangle$ denote an average. The velocity covariances may be regarded as generalizations of the Reynolds stresses, which are recovered for $\xi_k = 0$. Taylor (1938) also introduced the spectral representation of turbulence and reported on experiments showing that the turbulence behind a grid of cylinders in a wind tunnel is approximately isotropic far downstream.

The statistical theory of turbulence was the subject of particularly active developments in the 1940s and 1950s, primarily through works by a group in the USSR (Kolmogorov 1941; Millionschikov 1939, 1941; Obukhov 1941) and in Taylor's group in Cambridge (Batchelor 1953; Heisenberg 1948). Although this area of turbulence research has remained a very active one up until today, no completely deductive theory for isotropic turbulence has as yet come forward.

One area in which substantial progress has been made, however, is transition of a laminar flow to turbulence. The early investigation by Rayleigh concerning the stability of inviscid parallel shear flows was later followed by works in which the effects of viscosity were included. Orr (1907) and Sommerfeld (1908) independently derived the equation for infinitesimal wavelike perturbations of a parallel shear flow, which bears their names. The great difficulties in solving this, however, delayed further progress until about 1930, when useful solutions were worked out in Prandtl's group in Göttingen by Tietjens (1925), Tollmien (1929), and Schlichting (1933,

1935). It was long questioned whether the unstable so-called Tollmien–Schlichting waves could be observed in laboratory experiments. Early experiments to verify their existence generally proved unsuccessful. Schubauer and Skramstad (1947), who did experiments in which they introduced external periodic disturbances into the layer flow using an oscillating ribbon, were able to verify the linear stability theory in all its essential features. Later experiments have demonstrated that the linear instability is only an initial stage in a complex sequence of evolution of three-dimensional and nonlinear disturbances, which in its terminal phase leads to an abrupt breakdown of the flow into a fully developed turbulent boundary layer flow (see Klebanoff, Tidstrom, and Sargent 1962). For free-shear flows, as for jets and wakes, and for heat convection flows, the sequence of events appears to be more gradual.

The structure of fully developed turbulence in a shear flow has been studied in laboratory experiments by a large number of investigators. The development of the hot-wire anemometer has made such experiments possible. The very thorough investigation by Townsend summarized in two monographs (1956, 1976) should be especially mentioned and also the series of investigations by Klebanoff and his co-workers (Klebanoff 1954; Klebanoff and Diehl 1951; Schubauer and Klebanoff 1951; see also Laufer 1950, 1951 and 1954). In recent years the development of laser anemometry has made it possible to measure the instantaneous velocity field without introducing flow disturbances due to probes and their supports.

One particular feature of turbulence that has received a great deal of attention during the last two decades is the strong intermittency observed in the near-wall region of turbulent boundary layers and channel flows. In visualization experiments with a low-speed turbulent boundary layer, Kline et al. (1967) observed that the turbulence manifested itself mainly in short periods of strong activity, "bursts", separated by longer periods of nearly laminar unsteady flow. Such visual observations may be regarded as a form of conditional sampling, in which the experimenter observes and registers when something "interesting" happens. The criterion of what may be an interesting event may be, for example, the appearance of local high velocities and strong accelerations. Various such criteria have later been applied for a number of conditional sampling methods used in quantitative measurements of intermittent events in turbulent shear flows, and considerable progress has been made in the mapping of the intermittent turbulence structure. However, much interpretation and establishment of cause-and-effect relationships of the complicated observed sequences still remains to be done before a valid theoretical model can be formulated.

2

Characteristic scales and nondimensional parameters

Before embarking on a study of turbulent flows, be it theoretical or experimental, it is desirable to have estimates of the ranges of length, time, and velocity scales of motion to be found in the flow. An estimate of the upper limit of the length scales to be expected may be had from the dimensions of the flow apparatus, or the body generating the flow disturbance. Let this typical dimension be L. With a characteristic velocity U for the flow, one then finds a characteristic overall time scale of

$$t_a = \frac{L}{U},\qquad(2.1)$$

which is the time it takes for a fluid element to be advected past the body by the velocity U. A second characteristic time may be found from an estimate of the time of viscous diffusion across a distance L. From dimensional reasoning one finds the viscous diffusion time to be

$$t_v = \frac{L^2}{\nu},\qquad(2.2)$$

where ν is the kinematic viscosity. The Reynolds number, Re, may be interpreted as the ratio between these two, namely

$$\text{Re} = \frac{t_v}{t_a}.\qquad(2.3)$$

Estimates of typical values of the Reynolds number in flows of geophysical and engineering interest using such ν values as 1.5×10^{-5} m^2/s (for air at standard pressure and temperature) or 1.1×10^{-6} m^2/s (for water at 18 °C) show that they are usually very large. For example, for an airplane with a wing chord of 5 m flying at a velocity of 100 m/s at sea level, the Reynolds number is Re $= 3.3 \times 10^7$; for the flow in a river of width 1 km with a velocity of 1 m/s, Re $= 0.9 \times 10^9$. One may thus expect that for flows of such length and velocity scales viscous diffusion is much too slow

6

to be of any dynamical importance. However, in turbulent flows viscosity nevertheless always plays an important role in the dissipation of kinetic energy to heat, which predominantly takes place in the small-scale eddies. In fact, turbulent flows may in many important respects be controlled by the smallest eddies and hence indirectly by viscosity.

A high Reynolds number allows disturbances to develop large velocity gradients locally in the flow before viscous diffusion will have time to smear them out. The kinetic energy available through the release of large velocity differences in local instabilities could then initiate turbulence. For a flow of low Reynolds number, on the other hand, viscous diffusion will prevent the appearance of high local shear, and turbulence therefore cannot be maintained. Shear flow turbulence is thus a high-Reynolds-number phenomenon.

A nondimensional parameter playing somewhat the same role in thermally driven turbulence as the Reynolds number does in shear flow turbulence may be derived from simple reasoning as follows: Consider the motion of a fluid between two horizontal walls, the lower of which is heated at a temperature ΔT higher than the upper wall. The heating will cause thermal expansion of the fluid and therefore a lowered density by the amount $\Delta\rho$, which may be expressed in terms of a thermal expansion coefficient α as

$$\frac{\Delta\rho}{\rho} = -\alpha\,\Delta T. \tag{2.4}$$

A fluid element will therefore experience a buoyancy force per unit mass of $g\,\Delta\rho/\rho$, where g is the acceleration of gravity, and if allowed to accelerate freely from the lower to the upper wall, it would reach a convection velocity w_c' given by

$$w_c' \approx \sqrt{\text{acceleration} \times \text{distance}}\,,$$

so we choose

$$w_c' = \sqrt{g\alpha\,\Delta T\,d}\,. \tag{2.5}$$

From this, one can define a heat convection time,

$$t_c = \frac{d}{w_c'} = \sqrt{\frac{d}{g\alpha\,\Delta T}}\,. \tag{2.6}$$

The convection time may be compared to the heat diffusion time

$$t_h = \frac{d^2}{\kappa_h}, \tag{2.7}$$

where κ_h is the coefficient of thermal diffusivity

$$\kappa_h = \frac{k}{\rho c_p},$$

(2.8)

k being the coefficient of heat conduction and c_p the specific heat at constant pressure. The ratio t_h/t_c multiplied by the Reynolds number based on w_c' and d forms the nondimensional Rayleigh number,

$$\mathrm{Ra} = \frac{t_v t_h}{t_c^2} = \frac{g\alpha\,\Delta T d^3}{\kappa_h \nu},$$

(2.9)

an important quantity whose inverse gives a measure of how fast viscous and heat diffusion will smooth perturbations in the fluid caused by convection. If the Rayleigh number is large, the diffusivity is not sufficient to prevent the appearance of local density differences. Buoyancy forces associated with such density differences may drive a convective motion, which first is laminar and well ordered but with increasing Rayleigh numbers goes into more complicated and unsteady motion, eventually becoming turbulent at Ra values of 10^5–10^7.

Characteristic scales determined from overall length and velocity characteristics of the flow configuration as those above give but limited information of the typical scales found in turbulent flows. By making use of measured mean-flow quantities, such as the mean wall shear stress, one can define a wall friction velocity,

$$u_* = \sqrt{\frac{\sigma_w}{\rho}}$$

(2.10)

and a viscous wall length,

$$l_* = \frac{\nu}{u_*},$$

(2.11)

and a corresponding time scale,

$$t_* = \frac{l_*}{u_*} = \frac{\nu}{u_*^2}.$$

(2.12)

As will be shown in Chapter 5, these wall-related characteristic scales are useful for the description of the mean-flow quantities in the wall region of a turbulent boundary layer. They also give a measure of some of the turbulence characteristics in this region. Thus, the velocity fluctuation amplitudes are typically a few u_*, and the length scale of the smallest dynamically significant eddy may be a few l_*.

In a flow with density stratification the distribution of the mean velocity $\bar{U}(z)$ and the mean density distribution $\bar{\rho}(z)$ may be used to construct meaningful characteristic flow scales. For flows with stratification we will follow the custom in geophysical fluid mechanics and use z as the vertical coordinate. For pure shear flows, such as boundary layers, we will instead use y, as is customary in the aerodynamic literature. For a flow with density stratification the mean-flow rotation period

$$t_r = \left(\frac{d\bar{U}}{dz}\right)^{-1} \tag{2.13}$$

may be compared with the intrinsic oscillation period for the fluid,

$$t_s = \left(-\frac{g}{\bar{\rho}}\frac{d\bar{\rho}}{dz}\right)^{-1/2} = N^{-1}, \tag{2.14}$$

where N is known as the Brunt–Väisälä (or intrinsic) frequency. The square of the ratio of time scales is the Richardson number,

$$\text{Ri} = \left(\frac{t_r}{t_s}\right)^2 = \frac{-(g/\bar{\rho})(d\bar{\rho}/dz)}{(d\bar{U}/dz)^2}. \tag{2.15}$$

This nondimensional number plays an important role in the turbulence of a stratified shear flow. Large values of Ri (i.e., greater than unity) indicate strong static stability of the fluid and hence inhibition of turbulence, whereas for small values (less than a quarter) shear-induced instability and turbulence may set in.

It is often of interest to determine the smallest eddy scale that may be expected in a given turbulent flow. Dimensional arguments can be used to show that dissipation of mechanical energy into heat is caused primarily by the smallest eddies. Thus, the overall viscous dissipation rate per unit mass, ϵ, should be a useful quantity to which one should try to relate the smallest length scale. One can easily determine ϵ for a given stationary turbulent flow from the power required to drive the flow or in a flow in which no mechanical energy is added (such as in isotropic and homogeneous turbulence) by measuring the rate of decay of the fluctuations. Taking a typical small eddy scale to be l' and its velocity u', one finds that the dissipation rate is of the order

$$\epsilon \approx \nu \left(\frac{u'}{l'}\right)^2, \tag{2.16}$$

since the viscous stresses are proportional to the velocity gradients. The smallest eddy that survives long enough to be identified is the one for which the viscous diffusion time is about equal to the eddy advection time l'/u', that is, one for which its Reynolds number is about equal to unity, namely for

$$\mathrm{Re}' = \frac{u'l'}{\nu} \approx 1. \tag{2.17}$$

Combination of (2.16) and (2.17) gives for l' the Kolmogorov length scale

$$l_K = \eta = \left(\frac{\nu^3}{\epsilon}\right)^{1/4}, \tag{2.18}$$

and the corresponding time and velocity scales

$$t_K = \left(\frac{\nu}{\epsilon}\right)^{1/2} \tag{2.19}$$

and

$$v_K = (\nu\epsilon)^{1/4}, \tag{2.20}$$

respectively. A numerical example may be given as an illustration. Consider the mixing of 1 liter of water by an ordinary household mixer with an effective power of 10 W. At equilibrium conditions then

$$\epsilon = 10 \text{ W/kg},$$

and with $\nu = 10^{-6}$ m²/s (water at 20 °C) the Kolmogorov length scale is found to be

$$\eta = 2 \times 10^{-2} \text{ mm}.$$

In many engineering applications one is interested in achieving effective small-scale mixing, and estimates based on Kolmogorov's length scale may then be used to determine the smallest mixed eddy. Formula (2.18) then gives the rather discouraging information that by good mixing one pays a high price in power consumed; a halving of the smallest eddy size would require a power increase by a factor of 16.

A special length scale appearing in connection with atmospheric boundary layers is the Monin–Obukhov length scale,

$$L_{\mathrm{M-O}} = \frac{-\bar{\rho}_0 u_*^3 c_p}{\kappa \alpha g q}, \tag{2.21}$$

where $\bar{\rho}_0$ is the surface mean density, $\kappa = 0.41$ (von Karman's constant), and q is the vertical heat flux (positive upward). The Monin–Obukhov length (Monin and Obukhov 1953) gives a measure of the relative contribution of energy supplied to the turbulence by buoyancy forces to that

Table 1. *Characteristic scales*

	Length	Time	Velocity
General	l	t	U
Flow related	Body length L	Advection time $t_a = L/U$	U
Viscous	l_v	Diffusion time $t_v = l^2/\nu$	
Wall related	$l_* = \nu/u_*$	$t_* = \nu/u_*^2$	Friction velocity $u_* = \sqrt{\sigma_w/\rho}$
Turbulence related	Eddy scale l'	Eddy evolution time $t_c = l'/u'$	u'
	Kolmogorov: $l_K \equiv \eta = (\nu^3/\epsilon)^{1/4}$	$t_K = (\nu/\epsilon)^{1/2}$	$v_K = (\nu\epsilon)^{1/4}$
Thermal turbulence	Thickness of heated layer d	Convection time $t_c = \sqrt{d/g\alpha\,\Delta T}$	Convection velocity $w_c' = \sqrt{g\alpha\,\Delta T\,d}$
		Heat diffusion time $t_h = d^2/\kappa_h$	
Stratified shear flow		Mean-flow rotation period $t_r = (d\bar{U}/dz)^{-1}$	Mean velocity $\bar{U}(z)$
		Intrinsic oscillation period $t_s = (-g\bar{\rho}^{-1}\,d\bar{\rho}/dz)^{-1/2} \equiv N^{-1}$	
Atmospheric boundary layers	$L_{\text{M-O}} = -\bar{\rho}_0 u_*^3 c_p/\kappa\alpha gq$		

supplied by heat generated from friction; for large values of L_{M-O} the vertical heat flux has a small influence on the structure of the atmospheric boundary layer near the surface.

The different characteristic scales are summarized in Table 1 shown on the previous page.

3

Basic equations

The equations of motion for a continuum fluid express the conservation of mass, of momentum, and of energy. On the molecular scale a fluid has randomly fluctuating properties, and the continuum model represents an approximation in terms of averages over an ensemble or over small domains in space and time. The errors incurred in such an approximation depend on the ratio of the relevant spatial and temporal flow scales to those of the molecular fluctuations. In turbulence one also deals with flows that, in addition to a large-scale organized behavior, also contain random fluctuations. For such flows one finds it useful to construct models based on averages.

In this chapter we first develop the equations of motion for the special case of an incompressible continuum fluid. Then we express the flow field in terms of a deterministic field and a fluctuating field to obtain the equations of the form developed first by Reynolds (1894). Also, we add the Boussinesq (1897) approximation, useful for density stratified fluids, and we derive averaged equations for turbulent energy and vorticity.

Equations of motion

By comparing the estimated length scale of the smallest eddies in a turbulent flow field with typical intermolecular distances in the fluid, one finds that the latter are several orders of magnitude smaller than the former. Hence, the fluid may be safely modeled as a continuum. For a continuum with a velocity field U_i, density (ρ), and temperature T we have the following equations, expressed in a Cartesian coordinate system x_i:

Momentum:

$$\rho \frac{DU_i}{Dt} = \rho F_i + \frac{\partial \sigma_{ij}}{\partial x_j},$$

(3.1)

13

Continuity:

$$\frac{D\rho}{Dt} + \rho \frac{\partial U_i}{\partial x_i} = 0, \tag{3.2}$$

Energy:

$$T\frac{DS}{Dt} = (\sigma_{ij} + P\delta_{ij})\frac{\partial U_i}{\partial x_j} + \frac{\partial}{\partial x_j}\left(k\frac{\partial T}{\partial x_j}\right), \tag{3.3}$$

State:

$$P = f(\rho, T), \tag{3.4}$$

where

$$\frac{D}{Dt} = \frac{\partial}{\partial t} + U_j\frac{\partial}{\partial x_j},$$

t = time,

F_i = volume force,

σ_{ij} = stress tensor,

S = specific entropy,

k = coefficient of heat conduction,

P = pressure,

δ_{ij} = Kronecker delta.

For derivation of these equations, see, for example, Batchelor (1967).

In the following we shall consider a Newtonian incompressible fluid subjected to gravity forces and with small density variations caused by heating. Most of the interesting features of turbulence may be found in such fluids, and they also are commonly used in modeling of flows of geophysical interest. In the momentum equation we use the Newtonian stress-rate of strain relationship,

$$\sigma_{ij} = -P\delta_{ij} + 2\mu\left(e_{ij} - \frac{1}{3}\delta_{ij}\frac{\partial U_k}{\partial x_k}\right), \tag{3.5}$$

where μ is the coefficient of viscosity and e_{ij} the rate of deformation tensor,

$$e_{ij} = \frac{1}{2}\left(\frac{\partial U_i}{\partial x_j} + \frac{\partial U_j}{\partial x_i}\right) \tag{3.6}$$

combined with the equation of continuity (3.2) for an incompressible fluid,

$$\frac{\partial U_i}{\partial x_i} = 0. \tag{3.7}$$

The momentum equation then simplifies to

$$\rho \frac{DU_i}{Dt} = \rho F_i - \frac{\partial P}{\partial x_i} + \mu \frac{\partial^2 U_i}{\partial x_j \, \partial x_j}. \tag{3.8}$$

In the energy equation (3.3) the first term on the right represents the heat generated through deformation of fluid elements by viscous stresses. This contribution to the entropy production is generally unimportant for the low-speed flows that will concern us here. It will therefore be neglected. For the entropy we may set, with $S = S(P, T)$,

$$DS = \left(\frac{\partial S}{\partial P}\right)_T DP + \left(\frac{\partial S}{\partial T}\right)_P DT. \tag{3.9}$$

Neglecting the contribution from the pressure variations and setting

$$\frac{\partial S}{\partial T} = \frac{c_p}{T}, \tag{3.10}$$

where c_p is the specific heat at constant pressure, we then find from (3.3), upon neglect of the contribution due to viscous forces,

$$\frac{DT}{Dt} = \kappa_h \nabla^2 T, \tag{3.11}$$

where $\kappa_h = k/\rho c_p$ is the coefficient of thermal diffusivity.

For the equation of state we shall assume, in accordance with the assumption of an incompressible medium, that the density depends only on the temperature but not on the pressure. Thus, we set

$$\rho = \rho_0 [1 - \alpha(T - T_0)], \tag{3.12}$$

where subscript 0 denotes reference values and α is the coefficient of thermal expansion. Because $\alpha(T - T_0)$ generally is small in the class of problems of interest here, one may neglect the density variations and hence replace ρ by the constant value ρ_0, except in buoyancy terms. This leads to the *Boussinesq approximation*,

$$\rho \frac{DU_i}{Dt} = -\frac{\partial P}{\partial x_i} - g\rho[1 - \alpha(T - T_0)]\delta_{i3} + \mu \nabla^2 U_i, \tag{3.13}$$

$$\frac{\partial U_i}{\partial x_i} = 0, \tag{3.7}$$

$$\frac{DT}{Dt} = \kappa_h \nabla^2 T, \tag{3.11}$$

where we have omitted the subscript from ρ_0. (The continuity and energy equations are repeated for clarity.) These equations contain five unknowns, the three velocity components U_i, pressure P, and temperature T. The pressure may be formally eliminated by taking $\partial/\partial x_j$ of the ith component of (3.13) and subtracting $\partial/\partial x_i$ of the jth component [i.e., taking the curl of (3.13)], which gives, after some manipulations,

$$\rho\left(\frac{D\Omega_{ij}}{Dt} + \frac{\partial U_k}{\partial x_j}\Omega_{ik} - \frac{\partial U_k}{\partial x_i}\Omega_{jk}\right)$$

$$= g\rho\alpha\left(\frac{\partial T}{\partial x_j}\delta_{i3} - \frac{\partial T}{\partial x_i}\delta_{j3}\right) + \mu\nabla^2\Omega_{ij}, \tag{3.14}$$

where

$$\Omega_{ij} = \frac{\partial U_i}{\partial x_j} - \frac{\partial U_j}{\partial x_i}. \tag{3.15}$$

It is seen that Ω_{ij} is an antisymmetric tensor with only three independent components. Hence, we may replace it with a vector quantity, the vorticity ω_i, which may be formally related to Ω_{ij} by

$$\Omega_{ij} = -\epsilon_{ijk}\omega_k, \tag{3.16}$$

where ϵ_{ijk} is the alternating tensor (0 for any two indices being equal, $+1$ for any even number of permutations, and -1 for any odd number). By formally replacing by (3.16) in (3.14), one obtains an equation for ω_i. (This is actually most easily accomplished by considering one component at a time.) One finds, after some calculations,

$$\frac{D\omega_i}{Dt} = \omega_j\frac{\partial U_i}{\partial x_j} + g\alpha\epsilon_{i3}\frac{\partial T}{\partial x_j} + \nu\nabla^2\omega_i. \tag{3.17}$$

The first term on the right represents the contribution to the time rate of change due to vortex stretching and rotation, the second is the contribution from buoyancy forces, and the third the influence of viscous diffusion. Vortex stretching, which is basically a three-dimensional flow phenomenon, is generally considered the most important contributor to vorticity production in turbulent flows.

Equation (3.17) is also valid for a rotating coordinate system, provided one includes in the vorticity the contribution $2\Omega_i$ from the solid-body rotation with the angular velocity Ω_i.

The Boussinesq approximation neglects the cyclostrophic effect, namely that a horizontal pressure gradient applied to a stratified fluid will accelerate a light fluid more than a heavy one. The neglected term in (3.17) is the cyclostrophic vorticity generation term, namely,

$$\frac{\nabla \rho \times \nabla p}{\rho^2} = \nabla \times \left(-\frac{\nabla p}{\rho} \right).$$

Averaged equations

We consider the flow variables U_i, P, and T to consist of a mean part, which will be denoted by an overbar, and a fluctuating part with a zero average (by definition). The concept of statistical average will be discussed in more detail in Chapter 4; suffice it to say that here we will regard the average as an ensemble average obtained as the arithmetic mean over many identical experiments. (Hence, the averaged quantity may be time dependent.) Accordingly we introduce in the equation of motion

$$U_i = \bar{U}_i + u_i, \qquad P = \bar{P} + p, \qquad T = \bar{T} + \theta, \tag{3.18}$$

where, by definition, with $\langle \ \rangle$ denoting (ensemble) average,

$$\langle u_i \rangle = \langle p \rangle = \langle \theta \rangle = 0. \tag{3.19}$$

These are substituted into the equations of motion (3.13), (3.7), and (3.11) and averages taken. This yields, with the aid of the continuity equation, the following set of Reynolds-averaged equations:

$$\rho \left(\frac{\partial \bar{U}_i}{\partial t} + \bar{U}_j \frac{\partial \bar{U}_i}{\partial x_j} \right) = -\frac{\partial \bar{P}}{\partial x_i} - \rho g [1 - \alpha(\bar{T} - T_0)]\delta_{i3}$$

$$+ \frac{\partial}{\partial x_j} \left[\mu \frac{\partial \bar{U}_i}{\partial x_j} + \tau_{ij} \right], \tag{3.20}$$

$$\frac{\partial \bar{U}_i}{\partial x_i} = 0, \tag{3.21}$$

$$\rho c_p \left(\frac{\partial \bar{T}}{\partial t} + \bar{U}_j \frac{\partial \bar{T}}{\partial x_j} \right) = \frac{\partial}{\partial x_j} \left(k \frac{\partial \bar{T}}{\partial x_j} - q_j \right), \tag{3.22}$$

where

$$\tau_{ij} = -\rho \langle u_i u_j \rangle \tag{3.23}$$

is known as the Reynolds stress tensor and

$$q_i = \rho c_p \langle u_i \theta \rangle \tag{3.24}$$

is the turbulent (eddy) heat flux vector.

The process of obtaining (3.19) through (3.21) makes use of the following averaging rules for a quantity $U_i = \bar{U}_i + u_i$:

$$\langle u_i \bar{U}_j \rangle = \langle u_i \rangle \bar{U}_j = 0$$

and

$$\langle(\bar{U}_i+u_i)(\bar{U}_j+u_j)\rangle = \bar{U}_i\bar{U}_j+\langle u_iu_j\rangle$$

with corresponding expressions for terms involving derivatives.

The equations for the fluctuating components are found by subtracting the averaged equations from the full equations. This process yields

$$\rho\left(\frac{\partial u_i}{\partial t}+\bar{U}_j\frac{\partial u_i}{\partial x_j}+u_j\frac{\partial \bar{U}_i}{\partial x_j}\right)=-\frac{\partial p}{\partial x_i}+\rho g\alpha\theta\delta_{i3}$$

$$+\frac{\partial}{\partial x_j}\left(\mu\frac{\partial u_i}{\partial x_j}-\rho u_iu_j-\tau_{ij}\right), \qquad (3.25)$$

$$\frac{\partial u_i}{\partial x_i}=0, \qquad (3.26)$$

$$\rho c_p\left(\frac{\partial\theta}{\partial t}+\bar{U}_j\frac{\partial\theta}{\partial x_j}+u_j\frac{\partial\bar{T}}{\partial x_j}\right)=\frac{\partial}{\partial x_j}\left(k\frac{\partial\theta}{\partial x_j}-\rho c_pu_j\theta+q_j\right). \qquad (3.27)$$

An equation for the kinetic energy of the fluctuation velocities may be obtained by multiplying (3.25) by u_i, contracting the equations, and taking the average. One finds, after some manipulations with the aid of continuity, the following equation for the average kinetic energy $q=\frac{1}{2}\langle u_iu_i\rangle$ per unit mass:

$$\left(\frac{\partial}{\partial t}+\bar{U}_j\frac{\partial}{\partial x_j}\right)q=P+B-\epsilon-\frac{\partial F_i}{\partial x_i}, \qquad (3.28)$$

where,

$$P=-\langle u_iu_j\rangle\frac{\partial\bar{U}_i}{\partial x_j}$$

= production of turbulence/unit mass by the workings of Reynolds stresses against the mean shear,

$$B=g\alpha\langle u_3\theta\rangle$$

= production/unit mass through buoyancy,

$$\epsilon=\frac{\nu}{2}\left\langle\left(\frac{\partial u_i}{\partial x_j}+\frac{\partial u_j}{\partial x_i}\right)^2\right\rangle.$$

= viscous dissipation/unit mass,

$$\rho F_i=\langle pu_i\rangle-\mu\left\langle u_j\left(\frac{\partial u_i}{\partial x_j}+\frac{\partial u_j}{\partial x_i}\right)\right\rangle+\frac{\rho}{2}\langle u_iu_j^2\rangle$$

= transport of kinetic energy/unit area through the action of pressure, viscous stresses, and turbulent diffusion.

The last term serves to redistribute the kinetic energy. A uniform flow ($\partial \bar{U}_i / \partial x_j = 0$) without thermal heating will have no turbulence production, and the turbulence in such a flow must always eventually decay.

Stability and turbulent energy

The turbulent energy equation (3.28) illustrates how Reynolds shear stress can do work against the mean velocity shear and thereby transfer energy from the mean flow to the fluctuating field. A flow is unstable when a disturbance will grow with time. For the analysis of the initial stages of an instability, one can use a linear approximation to the perturbation equations. Then superposition of perturbations are permitted, and an arbitrary time-dependent perturbation may be represented by a superposition of sinusoids. For the linearized stability problem it is thus sufficient to consider sinusoidal disturbances, since an arbitrary disturbance may be constructed by superposition of Fourier modes. For a two-dimensional flow the work by the Reynolds stress component τ_{12} against the mean shear must be positive for a disturbance to grow with time.

Viscosity can play an important role in high-Reynolds-number turbulence, and it is important for both production and dissipation of the energy of fluctuations. It may appear from the energy equation (3.28) that an increase in the fluid viscosity will generally stabilize the flow, since the dissipation rate is proportional to the viscosity coefficient. However, because of the intricate way viscosity may act to change phase relationships in the flow, an increase in viscosity may actually also lead to destabilization. As an example, the flat-plate boundary layer has no inflection point in the flow and therefore should be inviscidly stable; nevertheless, it does become unstable due to the action of viscosity.

It is possible to understand such behavior by studying the redistribution of the total mechanical energy of the flow [not just how the "energy of the perturbation" given by the square of the fluctuation velocity changes locally with time, as considered in (3.28)]. When a disturbance develops in the flow, the kinetic and potential energy of the mean flow generally decreases, since the energy of the fluctuations has been taken from the mean flow. The net energy change due to a disturbance (sum of the energy of the perturbation and the change in the mean energy) may be locally either positive or negative. If it is positive, as it would be for a simple mass-spring-damper system, for example, an increase in the dissipation rate would lead to a lowering of the system energy. Hence, an increase in the friction would be stabilizing. If the total perturbation energy of the system is negative, however, an increase in the dissipation rate, and hence a lowering of the total kinetic and potential energy of the system, would cause an increase in the amplitude of the fluctuations in order to compensate for the decrease in the net energy. From studies of the stability of boundary layers over flexible surfaces it may be concluded (Landahl 1962) that boundary layer instability waves may constitute a system that is energy deficient in this sense and hence could be destabilized by viscosity.

Example: deep-ocean density inversions

In the deep ocean one sometimes finds density inversions bounding what may be layers of convective activity. Such density inversions can be supported by normal vertical Reynolds stresses in much the same manner as it is possible to keep a tennis ball indefinitely suspended in a mean position at some distance above the tennis racket by bouncing it with the racket in a periodic manner. The mean positions of the ball are at the average height, near the middle between the racket, and at the top position in the ball bounce. Going back to the formalism of the oceanic example, assuming zero mean velocity and taking the density inversion to be of magnitude $\Delta\rho/\rho = -\alpha\,\Delta T = 10^{-6}$, one finds, from (3.20),

$$g\alpha\,\Delta T = \frac{\partial}{\partial x_3}\langle u_3^2\rangle. \tag{3.29}$$

If the thickness of the inversion layer is $h = 10$ cm, the rms vertical velocity w is of order of magnitude

$$w = (\langle u_3^2\rangle)^{1/2} = \frac{gh\,\Delta\rho}{\rho},$$

or

$$u = O(10^{-1}\,\mathrm{cm/s}).$$

Inversions of the same kind supported by a normal Reynolds stress associated with convective turbulence are also present in the atmosphere.

4

Statistical tools for description of turbulence

In this chapter we introduce the concept of an ensemble average, which allows one to form averages for time-dependent processes. One such ensemble average statistical measure is the autocovariance (or autocorrelation) function. It gives information about the average time dependence of a process. The Fourier transform of the autocovariance, in turn, describes the frequency contents of the process. For two random functions of time one can define cross covariance between values of the two functions at different times. The Fourier transform of the cross covariance with respect to delay time gives the cross-spectral density. When these measures are independent of the choice of time origin, the processes are stationary. We shall look at examples of how one can derive a propagation speed from cross covariances or cross spectra and also see how one can find decay times and other properties of a random process. A useful application of spectra and covariance functions is to the relationship between input and output statistical measures for a linear system. From observations of the excitation and the response one is able to draw conclusions about the dynamics of a system. If one knows the system dynamics and some of the statistical properties of the input, one can find the statistical properties of the output, and vice versa.

Many fluid flows can be approximated by linear systems of equations. This means that, in turn, some flows may react to excitation in ways that we can analyze, especially if the excitation is weak.

An example of a linear response of a flow field to turbulence is the emission of acoustic waves from a turbulent jet, as first analyzed by Lighthill (1952) and discussed in Chapter 10. Flows that respond to excitation by divergent oscillation are unstable; such flows are discussed in Chapter 7.

Correlations and spectra depend upon the second moments of a joint probability density. In order to relate correlations and probability distributions, this chapter also outlines some of the elements of probability theory, including the central limit theorem and the normal distribution.

21

As an illustration of a non-normal distribution the log-normal distribution is presented.

Ensemble averages

A random function is a function that cannot be predicted from its past. An example of a random function of space and time is the velocity field in a turbulent jet. The macroscopic boundary conditions may be independent of time in the sense that the supply of air to the jet is as constant as one can manage it, but the velocity at a point varies in an unpredictable manner with time. The local time average velocity is different in different locations, as are other averages, for example, the average of the square of the velocity departure from the local mean, or the nth power of the velocity fluctuations. For flows that do not change with time, in the sense that the gross boundary conditions are constant, we can define time averages. These are useful parameters for description of the flow. For flows where the macroscopic boundary conditions change with time, time averages lose their usefulness, and we shall take recourse to ensemble averages as used in other problems in statistical mechanics.

Consider an ensemble of macroscopically identical experiments, each of which produces a record of a variable $u(t)$, where t is the time. The output u of the jth experiment we call the jth realization of $u(t)$ and denote it $^{j}u(t)$. We can picture the ensemble of realizations of u as shown in Figure 4.1. The u's for all the realizations look like damped oscillations with superimposed "noise." Both the noise and the oscillation may differ from realization to realization. The reasons they vary depend on the physics of the underlying process, about which we have not said anything.

Now denote the ensemble average of the values of u by $\langle u(t) \rangle$, and define it as the limit

$$\langle u(t) \rangle = \lim_{N \to \infty} \frac{1}{N} \sum_{j=1}^{N} {}^{j}u(t). \qquad (4.1)$$

One can also define the ensemble average of a function of u like $g(u)$ in the same way,

$$\langle g[u(t)] \rangle = \lim_{N \to \infty} \frac{1}{N} \sum_{j=1}^{N} g[{}^{j}u(t)]. \qquad (4.2)$$

The ensemble average of the powers of u are the moments, so that the ensemble average of u^r at time t is called the rth moment of u at t. The rth moment is defined as

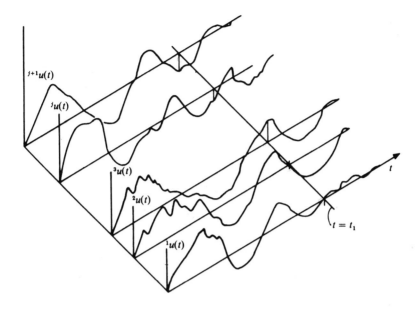

Figure 4.1. Ensemble of random functions of time, $^j u(t)$.

$$\langle u(t)^r \rangle = \lim_{N \to \infty} \frac{1}{N} \sum_{j=1}^{N} {}^j u(t)^r. \qquad (4.3)$$

Joint moment, covariance

Next, choose two times, t and t', and form the ensemble average of the product $u(t)u(t')$ for each realization, and call the joint moment the co-variance R_{uu}, defined by

$$R_{uu}(t, t') = \langle u(t)u(t') \rangle. \qquad (4.4)$$

Stationary random functions

A random function is called stationary if all its moments and joint moments are independent of the choice of time origin. For example, one should expect the flow in a turbulent jet to be stationary after it has settled down, and we avoid times close to the starting time. So if $t = 0$ is the starting time and T is the time constant of the starting transients, then for $t \gg T$, one should expect values of velocities and other variables measured in the jet to be stationary random functions.

The concept of a stationary random function represents a significant simplification, since then, for example, an average like $\langle u(t) \rangle$ will be independent of time, as will all the averages of functions of u, including the averages of the powers of u.

Time covariances of stationary random functions should be independent of the choice of time origin, but they will depend on the time difference $\tau = t - t'$,

$$R_{uu}(\tau) = \langle u(t)u(t+\tau) \rangle. \tag{4.5}$$

The double subscript indicates that the covariance is the covariance of u with u; the argument τ shows that the samples are taken a time interval τ apart.

Properties of the time autocovariances

First, we show that $R_{uu}(\tau)$ is an even function by writing the definition of R and shifting the time origin by an amount $-\tau$ to get successively,

$$R_{uu}(\tau) = \langle u(t)u(t+\tau) \rangle = \langle u(t'-\tau)u(t') \rangle = R_{uu}(-\tau). \tag{4.6}$$

This shows that R_{uu} is an even function of τ. Next, look at the cross covariance of u and its time derivative, $R_{u\dot{u}}(\tau)$, defined as

$$R_{u\dot{u}}(\tau) = \langle u(t)\dot{u}(t+\tau) \rangle = \frac{\partial}{\partial \tau} \langle u(t)u(t+\tau) \rangle = \frac{\partial}{\partial \tau} R_{uu}(\tau). \tag{4.7}$$

This can be generalized further, and we suggest that it will be useful to practice expressing covariances between derivatives in terms of derivatives of covariances.

Joint covariance

Next, consider joint (or cross) covariances. Take two joint random functions – that is, in each realization there are two results – a function $u(t)$ and a function $v(t)$. For example, they could be input and output of an electronic device, or, to choose a turbulence example, two velocity components at a point in the flow. The joint covariance function is defined as

$$R_{uv}(\tau) = \langle u(t)v(t+\tau) \rangle. \tag{4.8}$$

We have assumed the process to be stationary, since we have written the joint covariance as dependent on the relative time delay τ only.

The covariance function is not symmetric in time delay, rather, as one can easily show, generally, for a stationary process,

$$R_{uv}(\tau) = R_{vu}(-\tau). \tag{4.9}$$

For the covariance of u and the first time derivative of v one finds

$$R_{u\dot{v}}(\tau) = \frac{\partial}{\partial \tau} R_{uv}(\tau) = \frac{\partial}{\partial \tau} R_{vu}(-\tau). \tag{4.10}$$

Spectra and cross spectra

The covariance functions and cross covariance functions will be assumed to fall off for large values for the delay time, so that the functions are square integrable and possess Fourier transforms. The Fourier transforms of the time autocovariance function $R_{uv}(\tau)$ is called the power spectral density, defined by

$$S_{uu}(\omega) = \frac{1}{2\pi} \int_{-\infty}^{\infty} e^{i\omega\tau} R_{uu}(\tau) \, d\tau. \tag{4.11}$$

Fourier's integral theorem written for a function $f(t)$ reads

$$f(t) = \frac{1}{2\pi} \int_{-\infty}^{\infty} e^{-i\omega t} \int_{-\infty}^{\infty} e^{i\omega t'} f(t') \, dt' \, d\omega. \tag{4.12}$$

Applying it to the power spectral density, one has

$$R_{uu}(\tau) = \int_{-\infty}^{\infty} e^{-i\omega\tau} S_{uu}(\omega) \, d\omega. \tag{4.13}$$

The cross-spectral density of the joint pair of random functions u and v is defined as

$$S_{uv}(\omega) = \frac{1}{2\pi} \int_{-\infty}^{\infty} e^{i\omega\tau} R_{uv}(\tau) \, d\tau = \mathrm{Co}_{uv}(\omega) + i \, \mathrm{Qu}_{uv}(\omega). \tag{4.14}$$

The real part Co_{uv}, called the cospectrum, and the imaginary part Qu_{uv}, called the quadrature spectrum, may be expressed as integrals over positive τ as

$$\mathrm{Co}_{uv}(\omega) = \frac{1}{2\pi} \int_{0}^{\infty} \{[R_{uv}(\tau) + R_{uv}(-\tau)] \cos(\omega\tau)\} \, d\tau, \tag{4.15a}$$

and

$$\mathrm{Qu}_{uv}(\omega) = \frac{1}{2\pi} \int_{0}^{\infty} \{[R_{uv}(\tau) - R_{uv}(-\tau)] \sin(\omega\tau)\} \, d\tau. \tag{4.15b}$$

The cross-spectral density is also expressible in terms of its magnitude and argument. One usually divides the magnitude by the square root of the product of the spectral densities of the functions to write the cross-spectral density S_{uv} in terms of the coherence, $\text{Coh}_{uv}(\omega) = [(\text{Co}_{uv}^2 + \text{Qu}_{uv}^2)/S_{vv}S_{uu}]^{1/2}$, and phase, $\theta_{uv}(\omega) = \arg(S_{uv})$, as follows:

$$S_{uv}(\omega) = [|S_{uu}(\omega)||S_{vv}(\omega)|]^{1/2}\,\text{Coh}_{uv}(\omega)\exp[i\theta_{uv}(\omega)], \qquad (4.16)$$

which should be compared to (4.14).

Space–time covariances

Let $u(\mathbf{x}, t)$ be a random function of position $\mathbf{x}$ and time t. One can form the space–time covariance

$$R_{uu}(\mathbf{x}, \mathbf{x}', t, t') = \langle u(\mathbf{x}, t)u(\mathbf{x}', t)\rangle. \qquad (4.17)$$

If u is a stationary random function, R_{uu} is independent of the choice of time origin. If R_{uu} is independent of the choice of spatial origin, and the same holds for other statistical measures as well, one can say that $u(\mathbf{x}, t)$ is a *homogeneous* function of $\mathbf{x}$. So, for a stationary and homogeneous random function, the space–time covariance is

$$R_{uu}(\boldsymbol{\xi}, \tau) = \langle u(\mathbf{x}, t)u(\mathbf{x} + \boldsymbol{\xi}, t + \tau)\rangle. \qquad (4.18)$$

One can Fourier transform both with respect to space and time, and we shall go through the transformation step by step. First, the power spectral density of u is the Fourier transform of the time autocovariance $R_{uu}(0, \tau)$, and is equal to the local power spectral density

$$S_{uu}(\omega) = \frac{1}{2\pi}\int_{-\infty}^{\infty} e^{i\omega\tau}R_{uu}(0, \tau)\,d\tau. \qquad (4.19)$$

The wave number spectrum $\Phi(\mathbf{k})$ is defined by

$$\Phi(\mathbf{k}) = \frac{1}{(2\pi)^3}\int_{-\infty}^{\infty}\int_{-\infty}^{\infty}\int_{-\infty}^{\infty} \exp[-i(\mathbf{k}\cdot\boldsymbol{\xi})]R_{uu}(\boldsymbol{\xi}, 0)\,d\xi_1\,d\xi_2\,d\xi_3. \qquad (4.20)$$

Finally, the wave number–frequency spectrum is defined as

$$\Phi(\mathbf{k}, \omega) = \frac{1}{(2\pi)^4}$$

$$\times \int_{-\infty}^{\infty}\int_{-\infty}^{\infty}\int_{-\infty}^{\infty}\int_{-\infty}^{\infty} \exp[-i(\mathbf{k}\cdot\boldsymbol{\xi} - \omega\tau)]R_{uu}(\boldsymbol{\xi}, \tau)\,d\xi_1\,d\xi_2\,d\xi_3\,d\tau. \qquad (4.21)$$

For a two-dimensional $\mathbf{x}$ domain the integration with respect to ξ_3 is not included, and the power of $1/2\pi$ is 3.

Example

As an example we show how various spectral measures may be used to determine the phase velocity of a random wave system. The cross-spectral density may be obtained by measuring the output from two gauges placed a small distance l apart. Assuming that the power spectrum is the same at x and $\lambda x + l$, one can write

$$S_{uv}(\omega; l) = S_{uu}(\omega)\,\mathrm{Coh}_{uv}(\omega)\,\exp[i\theta_{uv}(\omega)]. \tag{4.22}$$

From this, one can determine the phase velocity of each frequency Fourier component as follows: The cross-spectral density, being a complex function, can be written in terms of its modulus and phase; the latter is

$$\theta_{uv}(\omega) = \arg[S_{uv}(\omega; l)]. \tag{4.23}$$

The phase gives the information we are looking for, since the phase gives the time interval δt between arrivals of a wave, which is proportional to the phase difference times the wave period divided by 2π, namely,

$$\delta t = \frac{(2\pi/\omega)\theta_{uv}(\omega)}{2\pi} = \frac{\theta_{uv}(\omega)}{\omega}. \tag{4.24}$$

The phase speed is equal to the ratio of the distance between the gauges to the time delay, so one finds for the phase velocity as a function of frequency

$$c(\omega) = \frac{l}{\delta t} = \frac{l\omega}{\theta_{uv}(\omega)}. \tag{4.25}$$

This assumes that the growth or decay between the two probes is negligible. The phase is the average phase for the frequency considered, and since the phase speed may be expressed as frequency divided by wave number, the average wave number for the frequency is given by

$$k = \frac{\omega}{c} = \frac{\theta_{uv}(\omega)}{l}. \tag{4.26}$$

If, on the other hand, for the particular case of surface waves, one uses both a surface slope transducer and a wave gauge, one may obtain the phase velocity from the functions

$$S_{uu}(\omega) \tag{4.27}$$

and, with $u' = \partial u/\partial x$, the slope spectrum is

$$S_{u'u'}(\omega) = k^2 S_{uu}(\omega). \tag{4.28}$$

Therefore, the mean square wave number is given by the ratio

$$\langle k^2 \rangle = \frac{S_{u'u'}(\omega)}{S_{uu}(\omega)}. \tag{4.29}$$

The phase speed estimate from this information is

$$c_{\text{slope}}^2 = \frac{\omega^2}{\langle k^2 \rangle}. \tag{4.30}$$

Note that the phase speeds determined from the cross-spectral density of the outputs of two wave gauges and the phase speed calculated from the spectrum and the slope spectrum may well be different. For example, for a standing wave field the outputs of the two wave gauges would be in phase, so the phase speed found from the cross-spectral density would be zero while the slope spectrum method would give a nonzero phase speed. For a wave field where there are waves propagating in one and only one direction, the two methods should give similar results.

Response of linear systems to random excitation

Observation of a turbulent velocity field involves using a transducer and an amplifier to transform the measured variable into an electric signal that can be recorded for analysis in digital or analog format. The output of the measuring system must be corrected for instrument system response. For the case of linear response it is relatively simple to relate statistical measures of the input and the output.

Many flow processes can also be described approximately in terms of a linear response to deliberate or ambient disturbances. The initial formation of surface waves on the water in the lee of a beach where the waves are excited by traveling pressure disturbance in the boundary layer (Phillips 1957) may be analyzed as a linear response process. Jet noise is another example of the excitation of a linearly propagating acoustic field excited by turbulent pressure fluctuations generated by nonlinear turbulent processes in the jet. A third example is the random pressure fluctuations in a turbulent boundary layer, which may be regarded as excited by the nonlinear turbulent bursting process and propagated as Tollmien–Schlichting waves (Landahl 1967; Bark 1975).

The early response of an initially laminar flow to disturbances propagating from upstream will be nearly linear. Instability is indicated by a response to a local disturbance, which grows as the disturbance propagates downstream. If the disturbances are sinusoidal (as from an oscillating ribbon, a typical laboratory situation in flow stability experiments), the unstable response will be exponentially growing in the early stages, for which the response is approximately linear.

We shall first remind the reader about how one calculates the response of a linear system to a specified excitation in terms of a Duhamel superposition integral. We then proceed to discuss the response to random input.

Impulse response and superposition

Let $v(t)$ be the input and $u(t)$ be the output of a linear system with time-invariant properties, so that u and v are related by

$$L_t[u(t)] = \sum a_n \frac{d^n}{dt^n} u(t) = v(t). \tag{4.31}$$

The coefficients a_n are constants, so the system has time-invariant properties. When the input is stationary in time, the output will also be stationary. We shall be concerned with outputs solely due to the input excitation $v(t)$, and we postulate that no remnants of response to other excitations may be present. We write the input v in terms of a Dirac delta function $\delta(t)$ defined by

$$\int_{-\infty}^{\infty} \delta(t-t')v(t')\,dt' = v(t) \tag{4.32}$$

for any well-behaved function $v(t)$. Let the response of the system to a unit impulse be $w(t)$, which satisfies

$$L_t[w(t)] = \delta(t). \tag{4.33}$$

A change of variables gives

$$L_t[w(t-t')] = \delta(t-t'). \tag{4.34}$$

Multiply both sides by $v(t')$ to get

$$L_t[v(t')w(t-t')] = v(t')\delta(t-t'). \tag{4.35}$$

Now integrate with respect to t' to obtain

$$L_t\left[\int_{-\infty}^{\infty} w(t-t')v(t')\,dt'\right] = v(t). \tag{4.36}$$

Here $v(t')$ has been put inside the differential operator, since the differentiation is with respect to t, and not with respect to t'.

Comparing equation (4.36) with (4.31), one can conclude that

$$u(t) = \int_{-\infty}^{\infty} w(t-t')v(t')\,dt'. \tag{4.37}$$

This is Duhamel's superposition integral. If one now takes the Fourier transform, one obtains

$$U(\omega) = \frac{1}{2\pi} \int_{-\infty}^{+\infty} e^{i\omega t} u(t)\, dt = W(\omega) V(\omega).$$ (4.38)

where $W(\omega)$ is the Fourier transform of $w(t)$ multiplied by 2π.

Next, consider random inputs.

Response to a random input

Let the input function $v(t)$ be a random, stationary function with zero mean value. We write (4.31), with a change in variable, as

$$L_\tau[u(t+\tau)] = v(t+\tau).$$ (4.39)

This takes advantage of the identity

$$\frac{d}{dt} = \frac{d(t+\tau)}{dt} \frac{d}{d(t+\tau)} = \frac{d}{d\tau}.$$ (4.40)

Multiply (4.39) by $u(t)$ and take the ensemble average. Since the ensemble averaging operation is independent of time t, the averaging operator can be put inside the operator $L_\tau[\ ]$, and one finds

$$L_\tau[\langle u(t)u(t+\tau)\rangle] = \langle u(t)v(t+\tau)\rangle,$$ (4.41)

or, in terms of the auto- and cross variances of u and v,

$$L_\tau[R_{uu}(\tau)] = R_{uv}(\tau).$$ (4.42)

If one instead multiplies (4.39) by $v(t)$, one finds

$$L_\tau[R_{vu}(\tau)] = R_{vv}(\tau).$$ (4.43)

These two equations are identical in form to (4.39), so their solutions may be expressed in the form of integrals, giving

$$R_{uu}(\tau) = \int_{-\infty}^{\infty} w(\tau - t') R_{uv}(t')\, dt',$$ (4.44)

$$R_{vu}(\tau) = \int_{-\infty}^{\infty} w(\tau - t') R_{vv}(t')\, dt'.$$ (4.45)

These two last equations are the Wiener–Khinchine relations.

Next, take the Fourier transforms of the last two equations to obtain the two algebraic equations for the spectra and cross spectra of input and output,

$$S_{uu}(\omega) = W(\omega)S_{uv}(\omega) \tag{4.46}$$

and

$$S_{vu}(\omega) = W(\omega)S_{vv}(\omega). \tag{4.47}$$

Note the order of the subscripts. Since

$$R_{uv}(\tau) = R_{vu}(-\tau),$$

one has that

$$S_{uv}(\omega) = S_{vu}^{*}(\omega). \tag{4.48}$$

where * denotes complex conjugate. (Note that $S_{vv} = S_{vv}^{*}$ since the auto-covariance function is an even function of τ.)

Combine (4.46) and (4.47) to obtain Campbell's theorem,

$$S_{uu}(\omega) = W(\omega)W^{*}(\omega)S_{vv}(\omega) = |W(\omega)|^2 S_{vv}(\omega). \tag{4.49}$$

This gives the output power spectral density in terms of a product of the frequency response function of the system and the power spectral density of the input. Now proceed to look at some examples.

Examples of response calculations

a. First-order system

For a simple first-order system such as a damped transducer, the typical response to input $v(t)$ is given by

$$\frac{d}{dt}u(t) + vu(t) = v(t). \tag{4.50}$$

The impulse response is

$$w(t) = H(t)e^{-vt}. \tag{4.51}$$

Taking the Fourier transform of the equation for the impulse response, one finds

$$W(\omega) = \frac{1}{v - i\omega}. \tag{4.52}$$

One is now ready to calculate the output for a given input. We suggest that the reader do so for an input that is a step function and for a square-wave periodic input.

The frequency response function $|W|^2$ relates the input and output spectra and also is equal to the output spectrum for white noise input, white noise having a power spectral density that equals a constant.

b. How little a spectral density tells about a signal

If one has succeeded in finding a prediction for the power spectral density of a flow variable, and the predicted power spectral density appears to be very much like the observed spectrum, one tends to feel a flush of triumph and momentarily deceive oneself into thinking that because the predicted spectrum matches the observation, the prediction must rest on correct physical insight. An absurdly simplified example will show that this is not necessarily so in all cases.

Consider a random signal that consists of unit impulses of random sign randomly spaced. The power spectral density of such a signal is a constant. Next, consider the signal one obtains by taking away every other impulse of the previous signal and multiplying the remaining impulses by $2^{1/2}$. The new signal will have the same power spectral density and the same mean square value. Now repeat this process many times, until the average time interval between impulses is 1×10^6 times the original average interval. The original signal could, for example, have been the force due to molecular impact on a membrane, and the final signal would be more like the force due to the impact of a small rock every few days. While both processes produce identical power spectra, they are obviously different. So spectra contain very limited information, and one needs to be cautious in drawing conclusions about the physics from spectral information. On the other hand, one usually uses power spectral methods to predict the response of a system with known response characteristics, and for such purposes, spectral methods can be very powerful. As is clear from the example, information about phase and relative time delays between events is missing from power spectra.

Distribution function and probability density

The distribution function $F(U)$ of a random variable is defined in terms of an ensemble average as follows:

$$F(U) = \langle H(U-u) \rangle = \Pr[u < U]. \tag{4.53}$$

Here, U is the variable in the same domain $-\infty < U < +\infty$. Each event or realization u can be marked off in the sample domain. We distinguish between the physical variable u and the coordinate U in the sample domain. The function $F(U)$ has the obvious properties

$$F(U) < 1 \quad \text{and} \quad F(U) \to 1 \quad \text{as} \quad U \to +\infty. \tag{4.54}$$

The probability density $f(U)$ (or simply "density function") is defined by

$$F(U) = \int_{-\infty}^{U} f(U')\,dU'. \tag{4.55}$$

For a continuous $F(U)$ this corresponds to

$$f(U) = \frac{dF}{dU} \equiv F'(U). \tag{4.56}$$

The average value $\langle u \rangle$ of u is related to the probability density function f by

$$\mu_1 = \langle u \rangle = \int_{-\infty}^{\infty} U f(U)\,dU. \tag{4.57}$$

This is the first moment of the density function $f(U)$. The nth moment μ_n is defined analogously as

$$\mu_n = \langle u^n \rangle = \int_{-\infty}^{\infty} U^n f(U)\,dU, \tag{4.58}$$

whereas for a function $h(u)$ of u one has

$$\langle h(u) \rangle = \int_{-\infty}^{\infty} h(U) f(U)\,dU. \tag{4.59}$$

This says that $\langle h(u) \rangle$ equals $h(U)$ times its likelihood of occurrence $f(U)\,dU$ summed over all U.

Characteristic function

The characteristic function of a random variable is defined from the probability density $f(U)$ by the Fourier integral

$$\phi(k) = \langle e^{ikU} \rangle = \int_{-\infty}^{\infty} e^{ikU} f(U)\,dU. \tag{4.60}$$

The inverse relation, from Fourier's theorem, is

$$f(U) = \frac{1}{2\pi} \int_{-\infty}^{\infty} e^{-ikU} \phi(k)\,dk. \tag{4.61}$$

The characteristic function can be calculated from the moments as follows: Expand the exponential in (4.60) to get, after one interchanges summation and integrations,

$$\phi(k) = \int_{-\infty}^{\infty} (1 + ikU - \tfrac{1}{2}k^2 U^2 + \cdots) f(U)\,dU$$

$$= 1 + ik\mu_1 - \tfrac{1}{2}k^2 \mu_2 + \cdots. \tag{4.62}$$

An example of a probability density often encountered is the normal, or Gaussian, distribution

$$f(U) = \left(\frac{1}{2\pi\sigma^2}\right)^{1/2} \exp\left(-\frac{U^2}{2\sigma^2}\right). \tag{4.63}$$

We have written it for zero mean, $\langle u \rangle = 0$. The only parameter needed to define $f(U)$ is the standard deviation σ. One finds that the odd moments are zero and the second moment is equal to σ^2.

One can calculate all the moments of the normal distribution from its characteristic function, which is

$$\phi(k) = (2\pi)^{1/2}\sigma \exp(-\tfrac{1}{2}\sigma^2 k^2). \tag{4.64}$$

For the details of this, and further information on this subject, the reader is referred to Cramér (1946).

Joint distributions, conditional probability

Consider a joint stochastic process of two variables, such as, for example, two components of a velocity vector $\mathbf{u}$ or the value of a velocity component in two different places in a velocity field. One can define a joint distribution function $F(U, V)$ in the (U, V) sample space as

$$F(U,V) = \langle H(U-u) \cdot H(V-v) \rangle = \Pr(u < U \text{ and } v < V)$$

$$= \int_{-\infty}^{U} \int_{-\infty}^{V} f(U', V') \, dU' \, dV'. \tag{4.65}$$

Here, $f(U, V)$ is the joint probability density.

The joint moments of the joint density function are given by

$$\mu_{mn} = \int_{-\infty}^{\infty} \int_{-\infty}^{\infty} U^m V^n f(U, V) \, dU \, dV. \tag{4.66}$$

The joint covariance is μ_{11}, and so on. The relationship between the probability densities $f(U)$, $f(V)$ and the joint distribution $f(U, V)$ is

$$f(U) = \int_{-\infty}^{\infty} f(U, V) \, dV, \tag{4.67}$$

and similarly for $f(V)$.

The conditional probability density, defined as the probability that $U < u < U + dU$ on the condition that $V < v < V + dV$, is written as

$$f(U|V) = \frac{f(U, V)}{f(V)}. \tag{4.68}$$

When u is *independent* of v, $f(U|V) = f(U)$, and one then finds

$$f(U)f(V) = f(U,V). \tag{4.69}$$

This says that the joint density of two independent variables is the product of the individual densities. In fact, calculating $f(U)f(V)$ and $f(U,V)$ from a set of observations, one can assess the degree of independence and measure it by calculating joint moments.

Central limit theorem

The central limit theorem states that under certain conditions probability distributions will tend to approach the normal distribution. The different versions of the central limit theorem differ in the conditions and restrictions imposed. For a full discussion and rigorous treatment Cramér (1946) is recommended reading. Here, we shall give a heuristic proof of the simplest version of the central limit theorem.

Let x_n ($n = 1, 2, 3, ..., N$) be a set of stochastic variables. The stochastic variable x_j, for example, is the random outcome of process number j, which produces the mean $\langle x_j \rangle$, a density function $f_j(X)$, and its associated characteristic function $\langle \exp(ikx_j) \rangle$, determined by ensemble averages over the realizations of process number j. Likewise, for all the other processes $n = 1, 2, ..., N$.

Thus, we shall deal with an ensemble of ensembles. The central limit theorem concerns the statistical properties of a new variable y, defined by

$$y = \frac{1}{N^{1/2}} \sum_{j=1}^{N} x_j. \tag{4.70}$$

The really significant and intellectually satisfactory import of the central limit theorem is that we can draw conclusions about the statistical properties of y even if one knows relatively little about the properties of the x's.

But we do have to restrict the properties of the x's quite severely. For the simplest form of the central limit theorem we specify that all the x's have identical probability densities with zero means and equal standard deviations σ. This definition keeps y from becoming infinite for infinite N. So we postulate that the x's have identical distributions and zero means and that the x's are all independent of one another.

Then, the mean square σ^2 of y is

$$\langle y^2 \rangle = \frac{1}{N} \sum_{j=1}^{N} \langle (x_j)^2 \rangle = \sigma^2, \tag{4.71}$$

since $\langle x_n x_m \rangle = \delta_{mn} \sigma^2$ because of independence. So now we know that

$$\langle y \rangle = 0 \quad \text{and} \quad \langle y^2 \rangle = \sigma^2. \tag{4.72}$$

The next step is to determine the characteristic function $\phi(k)$ for y defined by

$$\phi(k) = \int_{-\infty}^{\infty} e^{iky} f(y) \, dy = \langle e^{iky} \rangle. \tag{4.73}$$

(We no longer make the formal distinction between the stochastic variable y and the sample space coordinate.) Substitute for y to obtain

$$\phi(k) = \langle e^{iky} \rangle = \left\langle \exp\left[\frac{ik}{N^{1/2}} \sum_{n=1}^{N} x_n\right] \right\rangle. \tag{4.74}$$

But since the distributions for the x's are identical,

$$\phi(k) = \left\langle \left(\exp\frac{ikx_1}{N^{1/2}}\right)^N \right\rangle. \tag{4.75}$$

Expand the exponentials,

$$\phi(k) = \left\langle \left[1 + ikx_1 + \frac{(ikx_1)^2}{2} + \cdots\right]^N \right\rangle \approx \left(1 - \frac{k^2 \langle x^2 \rangle}{2N}\right)^N. \tag{4.76}$$

Take the limit as $N \to \infty$ to obtain

$$\phi(k) = \lim_{N \to \infty} \left(1 - \frac{k^2 \sigma^2}{2N}\right)^N = \exp\left(-\frac{k^2 \sigma^2}{2}\right). \tag{4.77}$$

Taking the inverse Fourier transform, one finds that y has a normal distribution in the limit as N becomes infinitely large,

$$f(Y) = \left(\frac{1}{2\pi\sigma^2}\right)^{1/2} \exp\left(-\frac{Y^2}{2\sigma^2}\right). \tag{4.78}$$

One finds many examples of stochastic variables whose values are determined by independent additive increments. The best-known example of such a variable may be the momentum of a molecule in a dilute gas. The x momentum of a molecule is its mass times the velocity component u. The x momentum at a given time is the vector sum of all the momentum increments caused by past collisions with other molecules. Assuming the increments in x momentum to be independent and to have zero mean value, one can, from the central limit theorem, conclude that u is normally distributed,

$$f(U) = \left(\frac{1}{2\pi kT}\right)^{1/2} \exp\left(-\frac{U^2}{2kT}\right), \tag{4.79}$$

where T is the temperature and k is Planck's constant.

Now, we go on and contend that the y and z momenta are independent of the x momentum, arguing that the increments in the three directions are independent. Hence, the y and z velocities also have Gaussian distributions. The joint distribution $f(U, V, W)$ for these independent variables is the product of their distributions,

$$f(U, V, W) = f(U)f(V)f(W)$$
$$= \text{const. } \exp\left(-\frac{U^2 + V^2 + W^2}{2kT}\right). \tag{4.80}$$

By introducing spherical coordinates in velocity space and setting $c^2 = U^2 + V^2 + W^2$, one finds

$$f(c) = 4\pi\left(\frac{1}{2\pi kT}\right)^{3/2} c^2 \exp\left(-\frac{c^2}{2kT}\right), \tag{4.81}$$

which is the well-known Maxwell–Boltzmann distribution. One can regard the Maxwell–Boltzmann distribution of molecular velocities in a dilute gas as a consequence of the independence of the succession of collisions experienced by a molecule. The three components are identically distributed because there is no preferred direction. The distributions have a zero mean because the gas is at rest in the mean in the chosen coordinate system.

One may next ask why velocities in a turbulent fluid do not have a normal distribution. One answer is that the momentum increments given a fluid particle at successive times are not independent. Since eddies and convergence regions, for example, tend to be coherent and interact, this eliminates the independence of interactions with other fluid particles experienced by a fluid particle.

But at some scales the local motions may be nearly independent. As a result, turbulent fields are not Gaussian, although they are not very different from having a Gaussian distribution. This difference is a crucial property of the dynamics of turbulence.

If we want to analyze the dynamics of turbulence, the non-Gaussian properties have to be included, but if we are only concerned with the effects of turbulence, we may, after experimental verification, find that the Gaussian distribution may be an adequate approximation.

As a final cautionary note, the approximation may be wrong by a large factor where the probability density is very small, far from the mean. While this does not show up in a striking manner in a plot, the rare events, far from the mean, are those that concern insurance companies and enter into the probability of failure. So, although such errors may appear small, they can nevertheless be very important.

The lognormal distribution

We next discuss an example of non-Gaussian behavior that gives the lognormal distribution. Consider a variable x that experiences a change Δx when there is a change Δy in the variable y, related by

$$\Delta x = x\,\Delta y. \tag{4.82}$$

Such relationships are similar to those that occur for conditional hydrodynamic instability, where the temporal growth rate of a disturbance measured by the variable x depends on a larger-scale flow parameter $(y - y_0)$. The latter could, for example, be the deviation from a critical parameter such as a Reynolds number. So the growth rate depends on x and y as

$$\frac{\partial x}{\partial t} = x(y - y_0). \tag{4.83}$$

This gives an exponential growth or decay rate of x depending on the sign and magnitude of the $y - y_0$. The variable x could stand for "small-scale kinetic energy", $q = \langle u^2 \rangle / 2$, while $y - y_0$ could be the departure from the mean larger-scale shear. If one also assumes that the Reynolds stress can be approximated by the expression Cq, one may write a production equation (3.28)

$$\frac{\partial q}{\partial t} \approx Cq\frac{\partial \bar{U}_1}{\partial x_3} \tag{4.84}$$

where C is an empirical constant. One then sees that, to this crude approximation, $\partial q / \partial t$ is proportional to q and its growth rate is controlled by $\partial \bar{U}_1 / \partial x_3$.

This justifies one's curiosity about the statistical properties of variables generated by similar processes. Let us return to the simple statement (4.82) written in differential form:

$$dx = (x - a)\,dy. \tag{4.85}$$

Integration gives

$$\ln(x - a) = y + \text{const.} \tag{4.86}$$

Now, if the increments dy are statistically independent and similarly distributed and have zero means, y will tend to be normally distributed according to the central limit theorem; that is,

$$f(y)\,dy = \frac{1}{\sigma\sqrt{2\pi}}\exp\left(-\frac{y^2}{2\sigma^2}\right)dy. \tag{4.87}$$

To find the density function for x, which we denote by f_i, one substitutes for y from (4.86) into (4.87) to get the density of x per unit length in the y domain. To find the density in the x domain, one has to multiply the result by the Jacobian dx/dy, defined by (4.85). The result is the distribution function

$$f_1(x) = \frac{1}{\sigma(x-a)\sqrt{2\pi}} \, \exp\left\{-\frac{[\log(x-a)-\text{const.}]^2}{2\sigma^2}\right\}. \qquad (4.88)$$

One can then verify that the integral of the density over the interval $-\infty < x < +\infty$ is unity, as implied in (4.54). The lognormal distribution occurs in bacteriological samples, in epidemiology, and in a large number of biological examples. It also seems to be a good approximation for some turbulence quantities. An example from turbulence is given by Dorman and Mollo-Christensen (1973), who found such a distribution to be approximately valid for the quantity $(\partial u/\partial t)^2$ in the turbulent boundary layer over water waves on a lake.

One important characteristic of the lognormal distribution is that the "tail" of the distribution is higher than for the normal distribution, so that events several standard deviations from the mean are more likely than for a normally distributed variable. An indication of a need to look more closely at the distribution function is when the most probable value is larger than the mean value. This kind of skewness often is the first sign of a tendency to a lognormal distribution. One finds this in the distribution function for chlorophyll in the upper ocean, for example. The underlying process requires an encounter between a phytoplankton particle and a nutrient, leading to an incremental increase in the size of phytoplankton. The larger the phytoplankton concentration, the more effective it is in consuming nutrient.

But, as with the normal distribution, the lognormal distribution is also an approximation, often quite close to observations, but still possibly wrong by a large factor for values far from the mean; in other words, these and other well-known distributions are poor predictors of catastrophe, which is defined as the occurrence of values far away from the mean value.

5

Flows that are homogeneous in more than one spatial dimension

We shall here discuss some properties of fields of turbulence that may be regarded as homogeneous in one or more space variables, that is, that have all space–time covariances invariant to shifts in spatial origin. Also, we will introduce the concept of isotropy, and discuss some of the consequences thereof in applications to turbulent flow fields. We will draw heavily on the material presented in Chapter 4 on statistical tools.

First, however, we shall consider the mean-flow properties of some simple, but important turbulent flow fields.

Some spatially homogeneous fields

1. Turbulent flow in a long, two-dimensional channel

Provided the length, L, of the channel (Figure 5.1) is much greater than its depth, d, and provided one keeps away from the neighborhood of the ends, the turbulent flow field can be expected to be homogeneous in x and z (but not in y, of course). The mean properties of the flow field, $\bar{U}$ and $\bar{P}$, are then such that $\bar{U}$ and $\partial \bar{P}/\partial x$ are functions of y only. If the driving pressure difference,

$$\Delta \bar{P} = \bar{P}_1 - \bar{P}_0$$

is independent of time, the flow is also statistically stationary.

For a fluid of constant density the Reynolds equations for the mean flow, (3.20), become

$$0 = -\frac{\partial \bar{P}}{\partial x} + \frac{\partial}{\partial y}\left(\mu \frac{\partial \bar{U}}{\partial y} + \tau_{12}\right), \tag{5.1}$$

$$0 = \frac{\partial \bar{P}}{\partial y} + \frac{\partial \tau_{22}}{\partial y}, \tag{5.2}$$

where

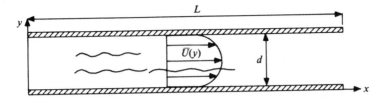

Figure 5.1. Turbulent flow in a channel.

$$\tau_{12} = -\rho\langle uv \rangle \qquad (u_1 = u, \ u_2 = v), \qquad (5.3)$$

$$\tau_{22} = -\rho\langle v^2 \rangle. \qquad (5.4)$$

All other terms, such as $\partial\tau_{11}/\partial x$ and $\partial\tau_{13}/\partial x$ in (5.1), drop out because of the spatial homogeneity in the x direction, which makes the x derivative of all mean quantities, with the exception of $\partial\bar{P}/\partial x$, equal to zero. One also finds that $\bar{V} = 0$, since, because of continuity,

$$\frac{\partial\bar{U}}{\partial x} + \frac{\partial\bar{V}}{\partial y} + \frac{\partial\bar{W}}{\partial z} = 0, \qquad (5.5)$$

and $\bar{U}$ and $\bar{W}$ are independent of x and z. By integration of (5.2), one obtains

$$\bar{P}(x,y) = \bar{P}_w(x) + \tau_{22}(x,y), \qquad (5.6)$$

where $\bar{P}_w$ is a linear function of x only. Substitution into (5.1) then gives

$$\frac{\partial}{\partial y}\left(\mu\frac{d\bar{U}}{dy} + \tau_{12}\right) = -G, \qquad (5.7)$$

where $G = -\partial\bar{P}_w/\partial x$ is a constant. Equation (5.7) may be integrated directly with the result

$$\mu\frac{d\bar{U}}{dy} + \tau_{12} = -Gy + \sigma_w, \qquad (5.8)$$

where $\sigma_w = \mu(d\bar{U}/dy)_w$ is the mean wall shear stress. A convenient way to represent (5.8) in a nondimensional manner is to introduce wall-related variables as in (2.10) and (2.11) and thus set

$$y^+ = \frac{yu_*}{\nu}, \qquad (5.9)$$

$$U^+ = \frac{\bar{U}}{u_*}, \qquad (5.10)$$

$$\tau^+ = \frac{\tau_{12}}{\rho u_*^2},\tag{5.11}$$

$$G^+ = \frac{G}{\rho \nu u_*^3},\tag{5.12}$$

where $u_* = \sqrt{\sigma_w/\rho}$ is the friction velocity. Then (5.8) takes the form

$$\frac{dU^+}{dy^+} + \tau^+ = -G^+ y^+ + 1.\tag{5.13}$$

Note that $\tau^+ = 0$ at the wall (u and v are zero there, hence $\langle uv \rangle = 0$) so that the nondimensionalization chosen automatically gives $dU^+/dy = 1$ at the wall. It is possible to relate G^+ to the depth of the channel by observing that the mean flow must be symmetric around the channel midpoint, which means that $dU^+/dy^+ = 1$ at $y = d$, that is,

$$G^+ = \frac{2}{d^+},\tag{5.14}$$

where $d^+ = du_*/\nu$. Hence, we have from (5.13)

$$\frac{dU^+}{dy^+} = 1 - \tau^+ - \frac{2y^+}{d^+}.\tag{5.15}$$

In order to calculate the velocity U^+, one needs to know the Reynolds shear stress, τ^+, which is not available from any exact theory without additional hypotheses regarding the relation between τ^+ and U^+. However, some conclusions may be drawn directly from (5.15). Typical values of d^+ for fully developed turbulent flows are very large, on the order of several thousands. Within the immediate neighborhood of the wall – for y^+ on the order of a few tens, say – one may therefore neglect y^+ compared to d^+, and (5.15) then gives

$$\frac{dU^+}{dy^+} + \tau^+ = 1.\tag{5.16}$$

Thus the sum of viscous and Reynolds stresses, the "total stress", is nearly independent of y^+ near the wall, where y^+ is small. According to (5.15) the total stress is really a linear function of y^+, but in the immediate neighborhood of the wall, the variations of dU^+/dy is dominated by τ^+ as compared to the term y^+/d^+. The main variation in dU^+/dy^+ occurs in the region close to the wall, so that τ^+ also varies rapidly there, as illustrated in Figure 5.2. Although the results shown in the figure were obtained for a boundary layer, in the immediate neighborhood of the wall they are equally valid for a channel flow.

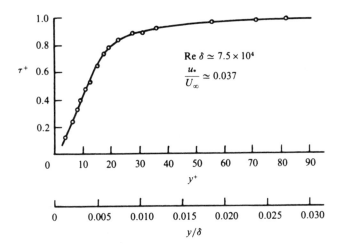

Figure 5.2. Distribution of turbulent shear stress in the wall region of a turbulent boundary layer (after Schubauer 1954).

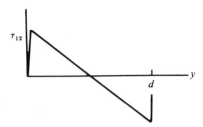

Figure 5.3. Variation of the Reynolds stress in a channel flow (conceptual).

Another interesting property becomes apparent when one looks at large values of y^+/d^+. Then, since $\bar{U}$ itself varies on a typical scale of d, dU^+/dy^+ becomes very small, and (5.15) may be approximated by

$$\tau^+ = 1 - \frac{2y^+}{d^+};\qquad(5.17)$$

that is, outside the region closest to the walls the Reynolds stress varies linearly with y, as illustrated in Figure 5.3.

2. Turbulent boundary layer

The flow in a turbulent boundary layer (Figure 5.4) differs from that in a channel in that the mean velocity and all statistical quantities change

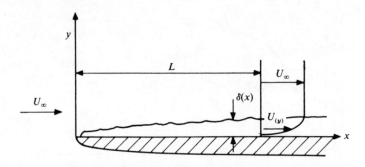

Figure 5.4. Turbulent boundary layer, notation.

slowly with downstream distance. Hence, there is no homogeneity in x for statistical quantities (but still homogeneity in z if the boundary layer is two dimensional). However, because of the slow streamwise variation of the mean-flow properties, homogeneity may still be invoked as a useful approximation for most of the statistical quantities. The Reynolds equations for the x and y components, under the assumption that the mean quantities vary only with x and y and the flow is steady and of constant density, read

$$\rho\left(\bar{U}\frac{\partial \bar{U}}{\partial x}+\bar{V}\frac{\partial \bar{U}}{\partial y}\right)=-\frac{\partial \bar{P}}{\partial x}+\frac{\partial}{\partial x}\left(\mu\frac{\partial \bar{U}}{\partial x}+\tau_{11}\right)+\frac{\partial}{\partial y}\left(\mu\frac{\partial \bar{U}}{\partial y}+\tau_{12}\right), \qquad (5.18)$$

$$\rho\left(\bar{U}\frac{\partial \bar{V}}{\partial x}+\bar{V}\frac{\partial \bar{V}}{\partial y}\right)=-\frac{\partial \bar{P}}{\partial y}+\frac{\partial}{\partial x}\left(\mu\frac{\partial \bar{V}}{\partial x}+\tau_{12}\right)+\frac{\partial}{\partial y}\left(\mu\frac{\partial \bar{V}}{\partial y}+\tau_{22}\right). \qquad (5.19)$$

To these is added the continuity equation,

$$\frac{\partial \bar{U}}{\partial x}+\frac{\partial \bar{V}}{\partial y}=0. \qquad (5.20)$$

As in the theory for laminar boundary layers one can bring out the most important terms by introducing a "stretched" boundary layer coordinate,

$$\eta=\frac{y}{\delta}, \qquad (5.21)$$

where $\delta=\delta(x)$ is the boundary layer thickness defined in some suitable way. By introducing this into the continuity equation, one finds that

$$\frac{\bar{V}}{\bar{U}}=O\left(\frac{\delta}{L}\right)\ll 1, \qquad (5.22)$$

where L is a typical streamwise dimension (for example, the distance from the leading edge). By substituting (5.22) into (5.18) and (5.17), assuming that η derivatives are of order unity, one finds that terms with y derivatives will generally dominate over those with x derivatives, and the terms that should be retained in (5.18) to be of lowest order in δ/L are thus found to be

$$\tilde{U}\frac{\partial \tilde{U}}{\partial \xi} + \tilde{V}\frac{\partial \tilde{U}}{\partial \eta} = -\frac{\partial \tilde{P}}{\partial \xi} + \frac{L}{\delta}\left(\mathrm{Re}_\delta^{-1}\frac{\partial^2 \tilde{U}}{\partial \eta^2} + \frac{\partial \tilde{\tau}_{12}}{\partial \eta}\right), \qquad (5.23)$$

where the following nondimensional quantities have been used:

$$\xi = \frac{x}{L}, \qquad \tilde{U} = \frac{\bar{U}}{U_\infty}, \qquad \tilde{V} = \frac{\bar{V}}{U_\infty}\frac{L}{\delta},$$

$$\mathrm{Re}_\delta = \frac{U_\infty \delta}{\nu}, \qquad \tilde{\tau}_{12} = \frac{\tau_{12}}{\rho U_\infty^2}, \qquad \tilde{P} = \frac{\bar{P}}{\rho U_\infty^2} \qquad (5.24)$$

and U_∞ is the free-stream velocity. It follows that balance of the terms, including the Reynolds shear stress term τ_{12}, requires that

$$\frac{\delta}{L} = O\left(\frac{\langle uv\rangle}{U_\infty^2}\right), \qquad (5.25)$$

which gives an estimate of the thickness of the turbulent boundary layer in terms of the fluctuation amplitudes. Typically, u and v are found to be on the order of one-tenth of the free-stream velocity so that the boundary layer thickness is on the order of 1% of the distance from the leading edge. Since Re_δ is generally much greater than unity, the viscous stress term in (5.23) may be neglected, except very close to the wall. The mean velocity profile therefore has the character illustrated in Figure 5.5 with a very steep gradient near the wall.

It is remarkable that on the basis of a very few and reasonable assumptions one can infer certain important properties of mean-flow distribution as demonstrated by Millikan (1939). First, a rescaling of (5.23) on the basis of wall-related variables yields (if terms involving $\partial\delta/\partial x^+$, which are small, are ignored)

$$\frac{1}{\delta^+}\left(U^+\frac{\partial U^+}{\partial x^+} + V^+\frac{\partial U^+}{\partial y^+}\right) = G^+ + \frac{\partial}{\partial y^+}\left(\frac{\partial U^+}{\partial y^+} + \tau_{12}^+\right), \qquad (5.26)$$

where $\delta^+ = \delta u_*/\nu$, $x^+ = x/\delta^+$, $y^+ = yu_*/\nu$, $U^+ = \bar{U}/u_*$, and $V^+ = \bar{V}/u_*\delta^+$. Since δ^+ generally is much greater than unity, we may neglect the left-hand side, which leads to the same equation as for the wall region of the two-dimensional channel flow. Thus, one would expect strong similarity

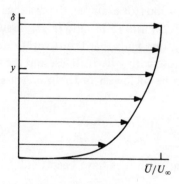

Figure 5.5. Typical turbulent boundary layer mean velocity distribution.

of all near-wall turbulent flows. Therefore, a *"wall law"* may be assumed of the form

$$U^+ = \frac{\bar{U}}{u_*} = f(y^+), \tag{5.27}$$

where f is a universal function depending only weakly on the wall-scaled pressure gradient G^+.

We next consider a representation valid near the edge of the boundary layer. The flow there will "see" the wall as generating a drag that will slow down the fluid and hence create a velocity defect, which may be represented by the *"velocity defect law"*,

$$\frac{U_\infty - \bar{U}}{u_*} = g(\eta). \tag{5.28}$$

The scaling with u_*, rather than with U_∞, say, is chosen because the magnitude of the velocity defect would depend on the wall friction and hence would be expected to be proportional to u_*. Next, following Millikan (1939), the assumption is made that the outer velocity defect law and the inner law overlap in some regions. From (5.27) and (5.28) we form

$$\frac{y\,\partial\bar{U}/\partial y}{u_*} = y^+ f'(y^+) = -\eta g'(\eta). \tag{5.29}$$

The two representations can only overlap if this function is a constant in some region of y. Thus, after integration

$$f(y^+) = \frac{1}{\kappa}\ln y^+ + A, \tag{5.30}$$

$$g(\eta) = -\frac{1}{\kappa} \ln \eta + B, \tag{5.31}$$

where $\kappa \approx 0.4$ is known as von Karman's constant. Clauser's (1956) observations gave values of $\kappa = 0.41$ and $A = 4.9$. Measured values of f and g are shown in semilog diagrams in Figures 5.6(a) and 5.7. In Figure 5.6(b) is shown results for a channel flow for comparison. It is seen that f has a logarithmic portion for y^+ greater than about 30. The extent of the logarithmic region ("logarithmic sublayer") depends on the overall flow parameters. Typically, it may occupy about one-tenth of the boundary layer thickness. The defect law, on the other hand, holds over most of the layer with the exception of the viscous sublayer, which generally is a very thin region, $y^+ = 30$, corresponding typically to about 1% of the boundary layer thickness at high Reynolds numbers. Coles (1956) has made extensive studies of the wall and velocity defect laws and has suggested the following uniformly valid representation of the velocity distribution:

$$\frac{\bar{U}}{u_*} = f(y^+) + \frac{\Pi w(\eta)}{\kappa}, \tag{5.32}$$

where

$$w(\eta) = \left[\frac{U_\infty}{u_*} - g(\eta) + \frac{1}{\kappa} \ln \eta - B \right] \frac{\kappa}{\Pi} \tag{5.33}$$

is Coles's wake function. This is found by adding the two representations of U and subtracting the expression for the overlap region. The constant Π in (5.32) depends on overall flow parameters and the upstream history of the flow. For a flat-plate boundary layer (i.e., zero pressure gradient) $\Pi = 0.55$. Coles (1956) found that the wave function may be well approximated for a wide class of flows by

$$w = 1 - \cos \pi \eta. \tag{5.34}$$

3. Thermal convection in a thin layer

A conceptually simple turbulent flow, which has been the subject of much study both experimentally and theoretically, is the flow that develops in a horizontally uniform layer heated from below when the temperature difference becomes sufficiently large.

Let us consider a layer of fluid of thickness d enclosed between two parallel, rigid, and perfectly conducting plates (Figure 5.8). The lower plate, located at $z = 0$, is held at a temperature T_0, whereas the upper one is held at $T_0 - \Delta T$. If the layer is thin compared to its horizontal dimensions,

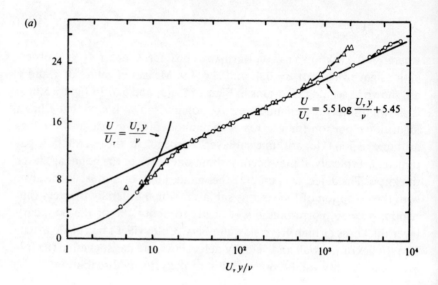

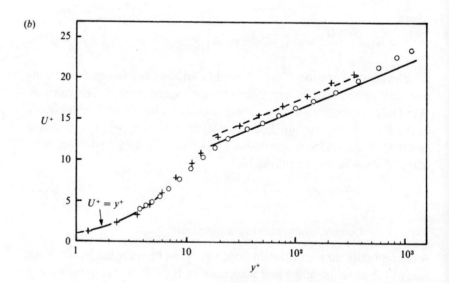

Figure 5.6. (*a*) Mean velocity distribution near smooth walls (from Patel 1965). (*b*) Mean velocity distribution in the wall and buffer regions of a channel flow. +, Re = 14,300; ○, Re = 54,000. —— and - - - - represent log laws with $\kappa = 0.41$ and $A = 5.6$ and 6.0, respectively (from Alfredsson and Johansson 1984; Johansson and Alfredsson 1982).

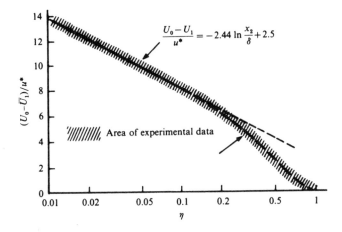

Figure 5.7. Velocity defect representation of the mean velocity distribution in a turbulent boundary layer.

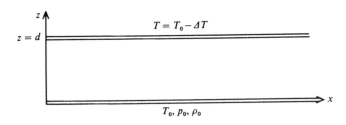

Figure 5.8. Fluid layer heated from below.

the turbulent velocity and temperature fluctuation fields may be considered statistically homogeneous in x and y (z is traditionally used as the vertical coordinate in this class of problem, as in geophysical fluid dynamics problems). The Reynolds equation (3.22) for the mean temperature distribution $\bar{T}(z)$ then becomes

$$\frac{d}{dz}\left(k\frac{d\bar{T}}{dz} - q_3\right) = 0, \qquad (5.35)$$

where q_3 is the vertical component of the turbulent (eddy) heat flux vector

$$q_3 = \rho c_p \langle w\theta \rangle. \qquad (5.36)$$

Integration of (5.35) gives

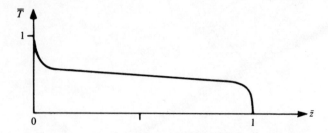

Figure 5.9. Typical mean temperature distribution for Nu ≫ 1.

$$q_3 - k\frac{d\bar{T}}{dz} = q_w, \tag{5.37}$$

where

$$q_w = -\left(\frac{d\bar{T}}{dz}\right)_w$$

is the mean heat flux at the wall. It is convenient to introduce nondimensional quantities by setting

$$\frac{q_w d}{k\,\Delta T} = \text{Nu}, \qquad \tilde{z} = \frac{z}{d}, \qquad \tilde{T} = \frac{T-T_0}{d}, \tag{5.38}$$

which gives for (5.37)

$$\frac{d\bar{T}}{d\tilde{z}} = \text{Nu}(\bar{q}-1). \tag{5.39}$$

The Nusselt number, Nu, is the ratio of the convective to conductive heat flux. Thus, Nu − 1 gives a measure of the relative heat transport due to turbulence. Generally, Nu ≫ 1 for well-developed convective turbulence. One would therefore expect the temperature distribution T to have large gradients very close to the wall, where $\bar{q}$ is small, but moderately small gradients in the inner portions of the fluid, where $\bar{q} \approx 1$ and the heat flux is dominated by the turbulent contribution (see Figure 5.9).

There will thus be thin conductive layers near each wall analogous to the viscous wall layers in wall-bounded turbulent shear flows. An important question is how the Nusselt number varies with overall flow parameters. One could get a rough idea of this by using for w the amplitude estimate given in Chapter 2. Under the assumption that the vertical velocity was produced by acceleration due to buoyancy, it was found that

$$w_c' \sim \sqrt{g\alpha\,\Delta T d}. \tag{2.5}$$

Thus, if the temperature fluctuations are of the order ΔT, the turbulent heat flux would be of order

$$\rho c_p \langle w'\theta \rangle \approx \rho c_p (\Delta T)^{3/2} d^{1/2} (g\alpha)^{1/2}. \tag{5.40}$$

Since this is close to the total heat flux, $q_w = \mathrm{Nu}(k\,\Delta T)/d$, one finds that

$$\mathrm{Nu} \sim \mathrm{Ra}^{1/2} \cdot \mathrm{Pr}^{1/2}, \tag{5.41}$$

where $\mathrm{Pr} = \nu/k$ is the Prandtl number. Actually, this gives an unsatisfactory estimate since experimentally $\mathrm{Nu}\cdot\mathrm{Ra}$ is found to have a piecewise linear variation with Ra between fairly distinct transition points at which the flow changes character (Malkus 1954a). Furthermore, for $\mathrm{Ra} \to \infty$ the Nusselt number becomes approximately proportional to $\mathrm{Ra}^{1/3}$. These properties will be further discussed in Chapter 11, where a more detailed description of turbulent convection will be presented.

Isotropic turbulence

Isotropy means invariance to orientation. In a discussion of isotropic turbulence homogeneity, that is, invariance to translation, is usually also understood.

First, let us consider the pressure–velocity covariances R_{p,u_i} ($R_{p,i}$ for short). Let the two points be located along the x axis. Then, by reflection about the x_1, x_3 plane

$$R_{p,2}(\xi_1) = -R_{p,2}(\xi_1), \tag{5.42}$$

which, because isotropy implies invariance to reflection and $R_{p,2}$ must be continuous at $\xi_1 = 0$, can only be true if

$$R_{p,2} = 0. \tag{5.43}$$

Similarly, by interchanging u_2 and u_3, one finds, again making use of invariance to reflection,

$$R_{p,3} = 0. \tag{5.44}$$

The only nonzero component is

$$R_{p,1} = \langle pu_1' \rangle, \tag{5.45}$$

where $u_1' = u_1(x_1 + \xi_1)$.

If we now consider an arbitrary separation ξ_k, it follows that $R_{p,i}$ must be a vector directed along the line connecting the two points; that is, one could set

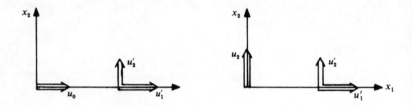

Figure 5.10. Longitudinal and transverse covariance components.

$$R_{p,i} = \xi_i R, \tag{5.46}$$

where R is a function only of the distance $r^2 = \xi_j \xi_j$ between the two points. Now we form the divergence of the vector $R_{p,i}$, namely,

$$\frac{\partial R_{p,i}}{\partial \xi_i} = \frac{\partial}{\partial \xi_i} \langle pu_i' \rangle = \left\langle p\frac{\partial u_i'}{\partial \xi_i} \right\rangle = \left\langle p\frac{\partial u_i'}{\partial x_i} \right\rangle = 0, \tag{5.47}$$

since only u_i' depends on ξ_i, and continuity must hold. Substituting (5.46) into (5.47), we obtain the equation

$$rR' + 3R = 0, \tag{5.48}$$

that is, $R \sim r^{-3}$, which means that it must be zero, since it cannot be infinite at the origin. Hence, all pressure–velocity covariances are zero in isotropic turbulence of an incompressible fluid.

Consider next the velocity–velocity covariances. Again, let the second point be along the x_1 axis (see Figure 5.10). By reflection in the x_1, x_3 plane we find that

$$R_{1,2} = -R_{1,2}.$$

Hence,

$$R_{1,2} = 0. \tag{5.49}$$

In a similar manner one finds that $R_{1,3}$ and $R_{2,3}$ are both zero. The only nonzero components are the *longitudinal covariance*

$$R_{1,1} = \langle u_1 u_1' \rangle = u_0^2 f(\xi_1), \tag{5.50}$$

where u_0 is the root-mean-square value of u_1 (which is the same for u_2 and u_3 because of isotropy), and the *transverse covariance*

$$R_{2,2} = R_{3,3} = u_0^2 g(\xi_1). \tag{5.51}$$

From these three expressions, all components of the tensor $R_{i,j}$ can be constructed for arbitrary separation ξ_k.

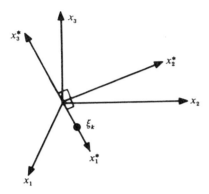

Figure 5.11. Rotation of coordinate system.

Introduce a new coordinate system (denoted by a star) rotated in such a way that its x_1^* axis passes through the point ξ_k (see Figure 5.11). One can express the velocity components u_i and u_j in terms of the components u_i^*, u_j^* in the rotated coordinate systems. One finds, simply,

$$u_j = e_{jk} u_k^*, \tag{5.52}$$

where e_{jk} is the directional cosine between the x_j axis and the x_k^* axis. From this, one can then form the covariances

$$R_{i,j} = e_{il} e_{jk} \langle u_k^* u_l^* \rangle = e_{il} e_{jk} R_{k,l}^*. \tag{5.53}$$

(Note the double sum over indices k and l.)

For $R_{k,l}^*$ one has, from (5.50) and (5.51),

$$R_{1,1}^* = u_0^2 f(r), \qquad r = \xi_k \xi_k,$$

$$R_{2,2}^* = R_{3,3}^* = u_0^2 g(r),$$

all the other components being zero. Therefore, one can write

$$R_{i,j} = e_{il} e_{jk} u_0^2 g(r) \delta_{kl} + u_0^2 [f(r) - g(r)] e_{i1} e_{j1}$$
$$= u_0^2 g(r) \delta_{ij} + u_0^2 [f(r) - g(r)] e_{i1} e_{j1}.$$

(Note that $e_{il} e_{jk} = \delta_{ij}$; try $x_i x_i = x_k^* x_k^* = r^2$.) But

$$e_{i1} = \frac{\xi_i}{r}, \qquad e_{j1} = \frac{\xi_j}{r}.$$

Hence,

$$R_{i,j} = u_0^2 \left\{ \frac{[f(r) - g(r)] \xi_i \xi_j}{r^2} + g(r) \delta_{ij} \right\}. \tag{5.54}$$

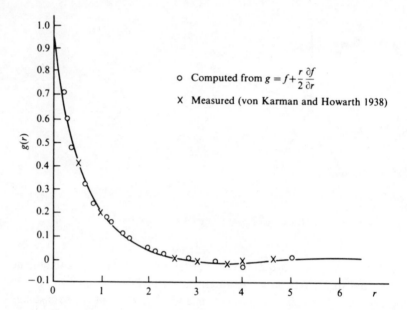

Figure 5.12. Transverse covariance $g(r)$ (from von Karman and Howarth 1938).

In a similar manner one can demonstrate that the triple covariance tensor

$$T_{ij,k} = u_i(x_l)u_j(x_l)u_k(x_l + \xi_l) \tag{5.55}$$

has only three independent components in a homogeneous and isotropic field and may be written

$$T_{ij,k} = u_0^3 \left\{ \frac{[k(r) - h(r) - 2q(r)]\xi_i \xi_j \xi_k}{r^3} \right.$$

$$\left. + \frac{h(r)\xi_k \delta_{ij}}{r} + \frac{q(r)\xi_j \delta_{ik}}{r} + q(r)\xi_i \delta_{jk} \right\}. \tag{5.56}$$

Continuity may be used to establish relationships between the different longitudinal and lateral covariance functions. Form the divergence

$$\frac{\partial R_{i,j}}{\partial \xi_j} = \left\langle u_i(x_k) \frac{\partial u_j(x_k + \xi_k)}{\partial x_j} \right\rangle = \left\langle u_i(x_k) \frac{\partial u_j'}{\partial x_j} \right\rangle = 0, \tag{5.57}$$

as follows from continuity. Upon introducing for $R_{i,j}$ the expression (5.54), one obtains the von Karman–Howarth relation

$$g(r) = \frac{rf'(r)}{2} + f(r) = \frac{1}{2r} \frac{\partial(r^2 f)}{\partial r}. \tag{5.58}$$

This may be used as a check on measured data (see Figures 5.12 and 5.13).

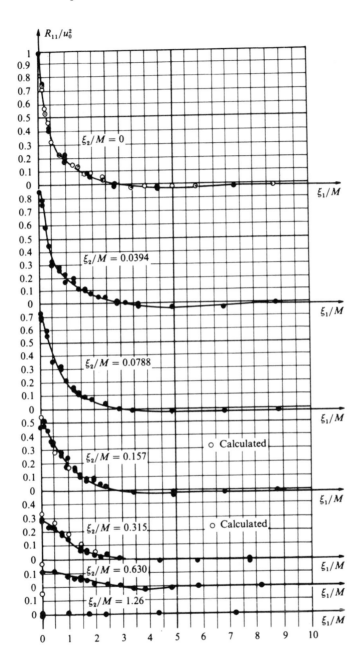

Figure 5.13. Spatial correlation coefficient as a function of ξ_1 and ξ_2 (M = mesh width) (from Favre, Gaviglio, and Dumas 1953).

In a similar manner one can show that

$$q(r) = -\frac{1}{2r}\frac{\partial}{\partial r}(hr^2), \tag{5.59}$$

$$k(r) = -2h(r). \tag{5.60}$$

Isotropic and homogeneous turbulence cannot be stationary. Consider the energy equation (3.28) for the isotropic and homogeneous case. For this, all spatial derivatives of mean quantities must be zero. Hence, in the absence of buoyancy,

$$\frac{\partial q}{\partial t} = -\epsilon = -\frac{\nu\langle(\partial u_i/\partial x_j + \partial u_j/\partial x_i)(\partial u_i/\partial x_j + \partial u_j/\partial x_i)\rangle}{2}, \tag{5.61}$$

so that the kinetic energy must always decrease with time. Isotropic turbulence is therefore *dying turbulence*. That it nevertheless is of considerable physical and practical interest is because observations show that in all turbulent flows the smallest eddies appear to behave as if they were isotropic. There is thus a *tendency toward isotropy* for the smallest scales. This is believed to be caused by the action of the pressure–velocity covariances, $R_{p,i}$, which are nonzero for nonisotropic fields.

In order to discuss the statistical behavior of the small-scale turbulence, it is convenient to apply a wave number decomposition to the covariances. We will hence work with the wave number spectra at zero time separation τ,

$$\Phi_{i,j} = \left(\frac{1}{2\pi}\right)^3 \int\int\int_{-\infty}^{\infty} R_{i,j}(\xi_l, t, 0) \exp(-ik_j\xi_j)\, d\xi_1\, d\xi_2\, d\xi_3. \tag{5.62}$$

A measure of the spectral distribution of the kinetic energy is provided by the trace of the tensor $\phi_{i,j}$, that is

$$\Phi_{i,i} = \left(\frac{1}{2\pi}\right)^3 \int\int\int_{-\infty}^{\infty} R_{i,i}(\xi_l, t, 0) \exp(-ik_j\xi_j)\, d\xi_1\, d\xi_2\, d\xi_3. \tag{5.63}$$

It should be remembered that the turbulence is not statistically stationary so that $R_{i,j}$ and hence $\phi_{i,j}$ are functions of the time t. By substituting the expression (5.54) for $R_{i,j}$, one finds

$$R_{i,i} = R_{1,1} + R_{2,2} + R_{3,3} = u_0^2[f(r) + 2g(r)] = u_0^2 R(r), \quad \text{say.} \tag{5.64}$$

Since $R_{i,i}$ depends on ξ_i through $r = \xi_i\xi_i$ only, one would expect that $\phi_{i,i}$ becomes a function of the magnitude of the wave number vector, $k = k_i k_i$, only but not of the individual components k_i. That this is indeed so may be demonstrated by the introduction of spherical-polar coordinate

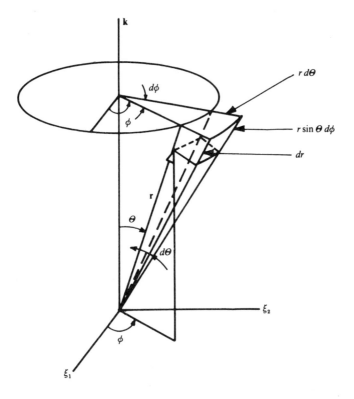

Figure 5.14. Spherical-polar coordinate system.

system r, ϕ, θ, oriented in such a manner that its polar axis is aligned with the k_i vector, north pointing in the positive k_i direction. The longitude is ϕ and θ is the latitude (see Figure 5.14).

We have

$$d\xi_1 \, d\xi_2 \, d\xi_3 = r^2 \sin\theta \, d\theta \, d\phi \, dr, \tag{5.65}$$

and

$$k_j \xi_j = \mathbf{k} \cdot \mathbf{r} = kr \cos\theta \tag{5.66}$$

(cf. the definition of a scalar product).

Thus,

$$\Phi_{i,i} = \frac{u_0^2}{(2\pi)^3} \int_0^\infty dr \int_0^{2\pi} d\phi \int_0^\pi r^2 R(r) \sin\theta \exp(ikr \cos\theta) \, d\theta$$

$$= \frac{2u_0^2}{(2\pi)^2} \int_0^\infty (kr)^{-1} \sin(kr) r^2 R(r) \, dr, \tag{5.67}$$

which indeed shows that $\Phi_{i,i}$ is a function of k only and not of k_1, k_2, and k_3 separately. It is convenient to define the *three-dimensional wave number (energy) spectrum*

$$E(k, t) = 2\pi k^2 \Phi_{i,i}(k, t). \tag{5.68}$$

This function measures how much energy is contained between the wave numbers k and $k + dk$. Namely, the inverse transform of (5.63) gives

$$R_{i,i}(r, t) = \iiint_{-\infty}^{\infty} \Phi_{i,i}(k, t) \exp(-ik_j \xi_j)\, dk_1\, dk_2\, dk_3$$

$$= 4\pi \int_0^{\infty} (kr)^{-1} \sin(kr) k^2 \Phi_{i,i}(k, t)\, dk, \tag{5.69}$$

where we have introduced a spherical-polar coordinate system in the k_i space in the same manner as for the ξ_i space above. Now, for zero spatial separation

$$R_{i,i}(0, t) = \langle u_i u_i \rangle = 4\pi \int_{\infty}^{0} k^2 \Phi_{i,i}(k, t)\, dk = 2 \int_0^{\infty} E(k, t)\, dk,$$

or, with $q = \langle u_i u_i \rangle / 2$,

$$q = \int_0^{\infty} E(k, t)\, dk. \tag{5.70}$$

The wave number spectrum E has been the object of much theoretical and experimental study. A convenient laboratory situation that approximates fairly well isotropic turbulence is the flow far downstream of a grid in a wind tunnel. In any point in such a flow the turbulence would be statistically stationary, hence strictly nonisotropic. However, because of the slow change with downstream distance, the turbulence would be nearly isotropic in a coordinate system moving with the mean velocity $\bar{U}$. Hence, time derivatives of the spectral quantities can be obtained approximately from

$$\frac{\partial}{\partial t} = \bar{U} \frac{\partial}{\partial x}. \tag{5.71}$$

This relationship also allows one to determine space covariances from time covariances under the assumption that the turbulent eddies are advected past the point of observation by the mean flow so fast that they do not have time to change substantially during their time of passage (Taylor's "frozen flow field hypothesis"). Results from wind tunnel measurement obtained by Van Atta and Chen (1969) far downstream of a grid ($x/M \gg 1$, where M is the mesh width of the grid) are shown in Figures 5.15a–d. The results are nondimensionalized by using the Kolmogorov scalings (2.18)–(2.20). It is seen from the data that, except when $\tilde{k} = k/k_K$ is very small, the measured spectra tend to be universal, that is, the non-dimensionalized spectra tend to be functions of $\tilde{k}$ alone. Here $k_\alpha = 2\pi/l_K$. The quantity

$$2\nu k^2 E(k) \tag{5.72}$$

measures the energy dissipation due to viscosity between the wave numbers k and $k + dk$. Let $T(k)$ denote the energy transferred between wave numbers k and $k + dk$ due to nonlinear effects. Then

$$\frac{\partial E}{\partial t} = T - 2\nu k^2 E. \tag{5.73}$$

One sees from the figures that $\partial E/\partial t$ tends to become small for k/k_K greater than about 0.3, so that an *equilibrium regime* is reached for high wave numbers. Kolmogorov (1941) hypothesized:

(i) For sufficiently high Reynolds numbers there is a region of high wave numbers for which the turbulence is (approximately) in statistical equilibrium and uniquely determined by the parameters ϵ and ν. This region is universal.

(ii) If the Reynolds number is infinite, there is a subrange such that

$$k_e \ll k \ll k_K, \tag{5.74}$$

where k_e is the wave number for the energy containing eddies (i.e., for the peak of E) in which E is independent of ν.

Dimensional analysis applied to the first hypothesis gives that

$$E = \alpha \epsilon^{2/3} k^{-5/3} E(\tilde{k}), \tag{5.75}$$

where α is a universal constant and $\tilde{k} = k/k_K$. Thus, from the second hypothesis when $\tilde{k} \ll 1$

$$E = \alpha \epsilon^{2/3} k^{-5/3}. \tag{5.76}$$

This is Kolmogorov's famous $k^{-5/3}$ law. It holds surprisingly well for spectra measured in high-Reynolds-number situations, as shown in Figure 5.16 obtained from data on atmospheric gusts.

The qualitative variation of E with k is illustrated in Figure 5.17. The universal equilibrium range extends in wave numbers from somewhere above the maximum of the spectrum and beyond. The shape in the lower k region depends on how the turbulence is created. Illustrated in this figure is the character of isotropic turbulence of very high Reynolds numbers. A relevant question is how one could define a Reynolds number for

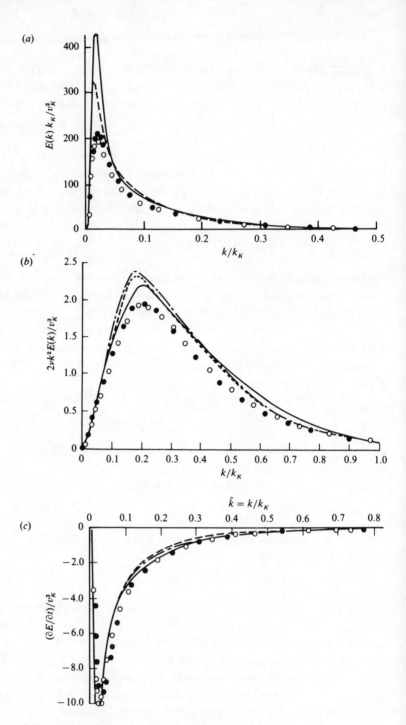

Figure 5.15. (a)–(c).

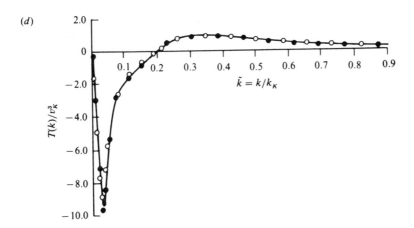

Figure 5.15. Normalized three-dimensional wave number spectra for isotropic turbulence (from Van Atta and Chen 1969). (a) Normalized three-dimensional energy spectra. •, $U = 15.7$ m/s; ○, $U = 7.7$ m/s. Uberoi (1954): ——, $x/M = 48$; - - -, $x/M = 72$. (b) Normalized three-dimensional dissipation spectra. •, $U = 15.7$ m/s; ○, $U = 7.7$ m/s. Uberoi (1954): ——, $x/M = 48$; - - -, $x/M = 72$; — · —, $x/M = 110$. (c) Rate of change of three-dimensional energy spectrum. •, $U = 15.7$ m/s; ○, $U = 7.7$ m/s. Uberoi (1954): ——, $x/M = 48$; - - -, $x/M = 72$; — · —, $x/M = 110$. (d) Measured three-dimensional energy transfer spectra. •, $U = 15.7$ m/s; ○, $U = 7.7$ m/s.

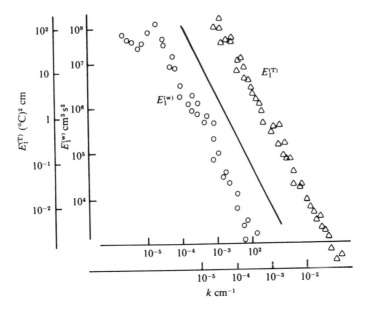

Figure 5.16. Spectra for temperature and vertical velocity fluctuations obtained from measurements from aircraft and from a tower (from Gurvich et al. 1967).

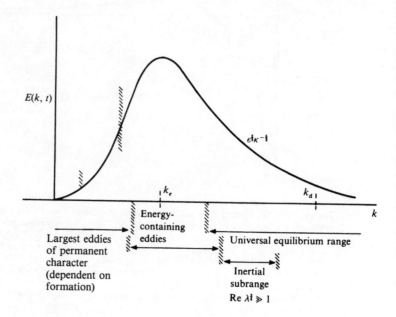

Figure 5.17. Behavior of the three-dimensional spectrum $E(k, t)$ in the various wave number ranges.

a turbulent flow that is supposed to be of universal character, that is, independent of the flow apparatus that created it. This can be done in different ways starting from either the longitudinal or the transverse correlation functions $f(r)$ or $g(r)$. Following Taylor, we shall use g.

An *integral scale*, Λ_g, may be defined as the first moment of g, that is,

$$\Lambda_g = \int_0^\infty g(r)\,dr. \tag{5.77}$$

This may be used as the characteristic length for forming a Reynolds number,

$$\mathrm{Re}_\Lambda = \frac{u_0 \Lambda_g}{\nu}. \tag{5.78}$$

It may be expected that Λ_g gives a measure of the size of the largest eddies in the flow. Also of interest is a measure of the size of the smallest dynamically significant eddies in the flow. This may be obtained by a Taylor series expansion of g around the origin,

$$g(r) = 1 + r\left(\frac{\partial g}{\partial r}\right)_0 + r^2\left(\frac{\partial^2 g}{\partial r^2}\right)_0 + \cdots.$$

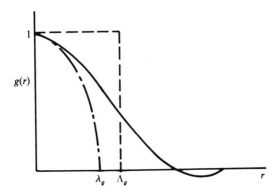

Figure 5.18. Taylor's micro- and integral scales.

The first term is zero because g must be an even function of r. Hence, we may set, approximately,

$$g(r) \approx 1 - \frac{r^2}{\lambda_g^2}, \tag{5.79}$$

where

$$\lambda_g = -\frac{2}{(\partial^2 g/\partial r^2)_0} \tag{5.80}$$

is *Taylor's microscale.* It may be thought of as the r intercept of the osculating parabola of $g(r)$ at the origin (see Figure 5.18). From this we form the Reynolds number

$$\mathrm{Re}_\eta = \frac{u_0 \lambda_g}{\nu}. \tag{5.81}$$

This has to be very much larger than unity for an initial subrange to exist.

6

Waves

Many of the flows that are studied in turbulence can support waves, i.e., nearly periodic disturbances that propagate relative to the fluid. Waves can be an important constituent of a turbulent field or even a dominant one, as in the case of wind-generated water surface waves. Such random wave fields can be described in the Fourier–Stieltjes sense as a superposition of waves of random phase and amplitude. Turbulent shear flows contain wave-like motions that give rise to statistical measures such as wave number–frequency spectra that are concentrated near dispersion lines in the wave number–frequency domain.

Transition to turbulence often occurs as wave-like bursts of local instability, and such instabilities can be both wave-like and wave-driven.

For flows that can support waves when the fluid is at rest, such as rotating flows, stratified flows, plasma flows, and compressible flows, waves will be an integral part of the field of turbulent fluctuations. Geophysical turbulence and plasma turbulence cannot be understood without taking account of the wave processes, and this holds for many other flows where wave motions play an important part in the dynamics of the flow.

We will here review some of the properties of waves and how they may be described, concentrating on aspects that are of specific importance in the study of turbulence.

Kinematic wave theory

Kinematic wave theory is a powerful tool for extending the results from the analysis of a uniform homogeneous wave train to more general wave-like disturbances. For a strictly spatially homogeneous case, as one would have, for example, for two-dimensional channel flow between infinite parallel walls, the solutions for waves of the form $\exp[i(\mathbf{k} \cdot \mathbf{x} - \Omega t)]$ could be employed together with Fourier superposition to generate disturbances of general spatial dependence. However, such superposition is not always

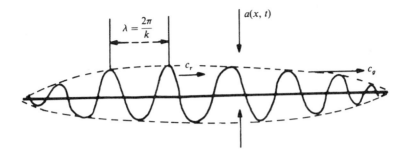

Figure 6.1. Slowly varying wave group.

transparent enough to produce straightforward and simple answers. A more direct way is to apply the results to wave groups with slowly varying wave number **k** and amplitude *a*, as illustrated in Figure 6.1.

The rate at which the wave number and amplitude vary along the wave group may be measured by a small nondimensional quantity ϵ ($\epsilon \ll 1$) where ϵ could be a measure of the relative change per wave length in amplitude and/or wave number. The number of waves in the whole wave group is of order ϵ^{-1}. The theory for wave groups with slowly varying properties, known as kinematic wave theory, may be deduced from an asymptotic expansion in ϵ (see Whitham 1974, Section 11.8). Here we shall give a short account of this theory as formulated for a one-dimensional wave group with $k = k(x, t)$. The extension to more than one dimension is easily carried out with the wave number and the group velocity treated as vector quantities. For a more detailed treatment of kinematic wave theory we refer to Whitham's (1974) book.

For an infinite and uniform linear wave train propagating through a uniform medium, the wave frequency is related to the wave number by the dispersion relation

$$\omega = i\gamma = \Omega(k) = \Omega^r + i\Omega^i, \qquad (6.1)$$

giving the frequency

$$\omega = \Omega^r(k) = kc_r, \qquad (6.2)$$

and the (temporal) growth rate

$$\gamma = \Omega^i(k) = kc_i \qquad (6.3)$$

for the wave train. We assume here that the waves have very small rates of growth or decay so that $|\Omega^i|/|\Omega^r| = O(\epsilon) \ll 1$. Now consider a wave

group with properties that vary slowly along the group. Its local frequency
and wave number may be defined from a phase function $\Theta(x, z, t)$, under
the assumption that the oscillatory disturbance varies like

$$a \exp(i\Theta), \tag{6.4}$$

where a is the amplitude, slowly varying with x and t, and Θ is assumed
to be real. From this phase function the wave number k and frequency ω
are defined as

$$k = \frac{\partial \Theta}{\partial x}, \tag{6.5}$$

$$\omega = -\frac{\partial \Theta}{\partial t}, \tag{6.6}$$

which are also assumed to be slowly varying with x and t. From these
definitions one obtains the compatibility condition

$$\frac{\partial k}{\partial t} + \frac{\partial \omega}{\partial x} = 0. \tag{6.7}$$

By introducing the dispersion relation, also assumed to have slow variations with x and t, one then finds

$$\frac{\partial k}{\partial t} + \Omega_k^r \frac{\partial k}{\partial x} = -\Omega_x^r, \tag{6.8}$$

which, after introduction of the definition of group velocity c_g,

$$c_g = \Omega_k^r, \tag{6.9}$$

may be written

$$\frac{Dk}{Dt} = -\Omega_x^r, \tag{6.10}$$

where

$$\frac{D}{dt} = \frac{\partial}{\partial t} + c_g \frac{\partial}{\partial x}. \tag{6.11}$$

In the same way one finds that

$$\frac{D\omega}{Dt} = \Omega_t^r, \tag{6.12}$$

In (6.8)–(6.10) subscripts denote partial derivatives in the extended space
(x, t, k).

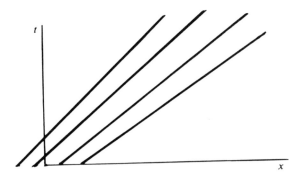

Figure 6.2. Trajectories for a homogeneous (one-dimensional) system of waves.

It follows from (6.8) and (6.10) that for a homogeneous system ($\Omega_k^r = 0$) the wave number is constant along the rays

$$\frac{dx}{dt} = c_g. \tag{6.13}$$

They are straight lines in the x, t plane since k (6.10) and c_g are constant along a ray (see Figure 6.2).

In addition to such kinematical relationships establishing how the wave number and frequency vary along the wave group, a dynamical equation is required to determine how the amplitude varies. The dynamics of slowly varying waves has been worked out by Whitham (1965) on the basis of the Lagrangian function, averaged over a cycle,

$$\bar{L} = \int_0^{2\pi} L \, d\Theta. \tag{6.14}$$

The average Lagrangian $\bar{L}$ (averaged kinetic energy minus averaged potential energy) is a quadratic functional of the amplitude, which is to be evaluated from the solution for an infinite homogeneous wave train. Whitham (1965) shows by using variational principles that for waves in a conservative system a conserved quantity is the wave action density A, defined by

$$A = \bar{L}_\omega. \tag{6.15}$$

For linear waves the flux of wave action is

$$B = c_g A, \tag{6.16}$$

so that wave action conservation implies

$$\frac{\partial A}{\partial t} + \frac{\partial(c_g A)}{\partial x} = 0.$$

(6.17)

For waves with a slight growth or decay this equation must be modified to (Jiminez and Whitham 1976; Chin 1980)

$$\frac{\partial A}{\partial t} + \frac{\partial(c_g A)}{\partial x} = 2\Omega^i A.$$

(6.18)

By integrating along the ray, one may write the solution of this as (Landahl 1972)

$$\frac{A}{A_0} = \frac{1}{J} \exp\left[2 \int_0^t \Omega^i dt\right],$$

(6.19)

where J is the Jacobian of ray position $x(t)$ with respect to the initial position $x(0) = x_0$, that is, J measures how the volume of a wave group changes along its trajectory.

In some situations the Jacobian may become zero somewhere along the trajectory, indicating space–time focusing of the wave group. At a focus the simple kinematic theory ceases to be valid, but one would expect that the amplitude could become large near such a point. For a steady one-dimensional wave train in a stationary background, one finds that the Jacobian is proportional to the group velocity c_g. For a group of short waves traveling on a background of steady, long waves with a phase velocity c_0, the background becomes steady in a frame of reference moving with the phase velocity so that a focus arises for $c_g = c_0$ for such a system (Landahl 1972).

For a steady-state background the frequency

$$\omega = \Omega^r = kc_r$$

(6.20)

is a constant along the ray. From this, one can determine how the wave number varies along the ray.

In a steady-state system it is often useful to translate the temporal rate of growth of the wave to a spatial growth rate. Ignoring the effect of the variation of J in (6.19) (which is zero for a spatially homogeneous system), and using the fact that wave action density is proportional to the square of the wave amplitude, one finds that, approximately,

$$a = a_0 \exp\left(\int_0^t \Omega^i dt\right) = a_0 \exp\left(\int_0^x k_i \, dx\right),$$

(6.21)

where

$$k_i = -\frac{k\Omega^i}{c_g}.$$

(6.22)

This is Gaster's (1962) transformation. It holds only for small growth rates $|\Omega^i|/|\Omega^r| \ll 1$ and for nonzero c_g. A wave with positive c_g and negative k_i is *convectively* unstable in the sense that an initial disturbance pulse will produce a wave group growing as it is convected downstream, but at any fixed point the disturbance will eventually decay. If, on the other hand, c_g is zero for some wave number within the unstable range the system is *absolutely* unstable in the sense that the disturbance will continue to grow everywhere (Bers 1975). For a wave propagating on a background inhomogeneity moving with a convection velocity c_0, Landahl's (1972) breakdown condition $c_g = c_0$ for an unstable disturbance thus implies that from a linear kinematic wave theory point of view the disturbance becomes absolutely unstable in a frame of reference moving with the velocity c_0.

Effects of weak nonlinearity

The linear kinematic wave theory may also be extended to weakly nonlinear waves by the addition to the dispersion relation of a term proportional to the square of the wave amplitude a (or proportional to A) by setting

$$\omega = \omega_{\text{linear}} + a^2 f_2(k) \equiv \Omega^r(k) + a^2 f_2(k) \tag{6.23}$$

(see Whitham 1974). The function $f_2(k)$ is to be determined from the dispersion relation for a uniform wave train of finite amplitude. As an example, f_2 is calculated below for a surface wave using Stokes's second-order solution. With this added term (6.10), (6.17), and (6.23) form a second-order system with the characteristics

$$\frac{dx}{dt} = \Omega_k^r \pm \sqrt{a^2 f_2 \Omega_{kk}^r}. \tag{6.24}$$

For $f_2 \Omega_{kk}^r > 0$ the characteristics are real. Because of the two families of characteristics, a given initial wave group will split into two with a splitting velocity proportional to the amplitude. If this produce is negative, on the other hand, the characteristics become complex, showing that the system is "partially elliptic". In such a case it is not possible to formulate a proper initial value problem for the wave train; initial sinusoidal perturbations of the wave amplitude will experience exponential growth (Whitham 1965), thus leading to instability and possibly eventual breakup of the wave group. This is also seen in the "sideband" instability of surface waves investigated by Benjamin and Feir (1967) and discussed further below. However, a proper treatment of such instabilities and their subsequent nonlinear development over a long time period requires the inclusion of additional terms in the equations for the evolution of the wave group.

In the study of the wave evolution over large times one has to include those effects that may have possible cumulative effects. This is the case with the nonlinear contribution to the frequency, as modeled in (6.23), which may lead to important changes over long times also for very small amplitudes. Kinematic wave theory may be insufficient for the study of long-time effects because it holds strictly only for times of order ϵ^{-1} times the basic wave period. Within this time the error in the kinematic theory is of order ϵ^2, provided $|\Omega^i|/|\Omega^r| = O(\epsilon)$. However, with the aid of the following two simple modifications the theory may be extended to hold also for times of order ϵ^{-2} times the basic period (see Chin 1980; Landahl 1982):

(i) Choose a complex wave number,

$$k = k^\kappa = k_r^\kappa + ik_i^\kappa$$

such that the group velocity becomes purely real, that is, such that

$$\Omega_k^i = 0;\tag{6.25}$$

the conservation equation (6.18) for the wave action is then changed to

$$\frac{\partial A}{\partial t} + \frac{\partial (c_g^\kappa A)}{\partial x} = 2(\Omega_\kappa^i - c_g^\kappa k_i^\kappa)A.\tag{6.26}$$

(ii) Modify the frequency relation (6.23) with an additional term to allow for the effects of rapid spatial amplitude variations as follows:

$$\omega = \Omega^r + f_2 a^2 - \frac{a_{xx}\Omega_{kk}^r}{2a},\tag{6.27}$$

all wave number derivatives to be calculated at the wave number k^κ.

A term equivalent to the third term in (6.27) was first derived by Chu and Mei (1970) for surface gravity waves using multiple scaling of the governing equations. This term is needed for cases when the amplitude variations are much greater than the phase variations, as could occur in the nonlinear long-time evolution of a dispersive wave packet. For a linear wave packet traveling in a uniform and steady medium it is not needed, however (see Landahl 1982).

One may expect that the dominant waves after a long time of travel are those with a wave number close to that for maximum growth rate, $k = k_m$. For that wave number the condition (6.25) is satisfied. Therefore, it is instructive to carry out a centered wave number expansion around $k = k_m$ for the complex function

$$\Psi = a(x, t)e^{i\Theta},\tag{6.28}$$

where Θ is the phase defined by

$$k - k_m = \Theta_k, \tag{6.29}$$

$$\omega - \Omega_m^r = -\Theta_t \tag{6.30}$$

(subscript m denoting quantities evaluated at $k = k_m$). After expanding (6.27) and (6.28) in $k - k_m$, retaining only the lowest-order terms, and setting A proportional to a^2 (without taking into account its variation with $k - k_m$), one finds, after some calculations, the following nonlinear equation:

$$\Psi_t + c_g \Psi_x - \tfrac{1}{2} \Omega_{kk} \Psi_{xx} = [\tfrac{1}{2}(\Omega^i - \Omega_{kx}^r) - i f_2 |\Psi|^2] \Psi; \tag{6.31}$$

all dispersion relation quantities are to be evaluated at $k = k_m$.

For a uniform medium ($\Omega_{kx}^r = 0$) this reduces to the Ginsburg–Landau equation arising naturally in many fluid dynamics problems and in the study of chemical reactions. For a conservative system it becomes the cubic nonlinear Schrödinger equation

$$\Psi_t + c_g \Psi_x - \tfrac{1}{2} i \Omega_{kk} \Psi_{xx} + i f_2 \Psi |\Psi|^2 = 0, \tag{6.32}$$

which has been derived in a number of different contexts with the use of different techniques.

Modulational instability

We study the stability of a train of finite-amplitude nondissipative waves by introducing into (6.32)

$$\Psi(x, t) = \Psi_0(t) + \Psi'(x, t), \tag{6.33}$$

where $|\Psi'|/|\Psi_0| \ll 1$. With Ψ_0 independent of x the solution is easily found from (6.32) to be

$$\Psi_0 = a_0 \exp(-i\sigma t), \tag{6.34}$$

where a_0 is the initial amplitude and

$$\sigma = f_2 a_0^2.$$

The oscillation in phase expressed by (6.34) is, of course, a consequence of the influence of finite amplitude on the frequency. Substitution of (6.33) and (6.34) into (6.32) and neglect of the quadratic terms in Ψ' then gives the following equation for Ψ':

$$\Psi_t' + c_g \Psi_x' - \tfrac{1}{2} i \Omega_{kk} \Psi_{xx}' + i f_2 (2\Psi' |\Psi_0|^2 + \Psi_0^2 \Psi'^*) = 0, \tag{6.35}$$

where the asterisk denotes complex conjugate and terms of order $|\Psi'|^2$ have been neglected. Following Benjamin and Feir (1967) we seek a solution for neighboring sideband wave numbers K and $-K$ of the form (see Whitham 1974, Section 15.6)

$$\Psi' = A_-(t) \exp[i(-K\xi + \omega_- t)] + A_+(t) \exp[i(K\xi + \omega_+ t)], \qquad (6.36)$$

where

$$\xi = x - c_g t$$

and ω_+ and ω_- are the corresponding sideband frequencies. They satisfy approximately the linear dispersion relation; hence

$$\omega_- = \omega_+ = \tfrac{1}{2} K^2 \Omega_{kk} = \bar{\omega}, \quad \text{say.} \qquad (6.37)$$

Substituting (6.36) into (6.35) and equating terms proportional to $\exp(-iK\xi)$ and $\exp(iK\xi)$ separately, one finds the following equations for the amplitudes A_+ and A_-:

$$\frac{dA_-}{dt} + i\sigma\{2A_- + A_+^* \exp[2i(\bar{\omega} - \sigma)]\} = 0,$$

$$\frac{dA_+}{dt} + i\sigma\{2A_+ + A_-^* \exp[2i(\bar{\omega} - \sigma)]\} = 0.$$

These have solutions of the form

$$A_\pm = a_\pm \exp(i\lambda_\pm t), \qquad (6.38)$$

where $\lambda_\pm$ satisfy

$$\lambda^2 + 2(\sigma - \bar{\omega})\lambda + \sigma(\sigma - 4\bar{\omega}) = 0, \qquad (6.39)$$

which has the solutions

$$\lambda_\pm = \bar{\omega} - \sigma \pm \sqrt{\bar{\omega}^2 + 2\bar{\omega}\sigma}. \qquad (6.40)$$

Sideband perturbations will grow if the roots are complex, that is, for

$$\bar{\omega}(\bar{\omega} + 2\sigma) < 0, \qquad (6.41)$$

which can only occur for $\bar{\omega}\sigma < 0$, that is, for

$$f_2 \Omega_{kk} < 0, \qquad (6.42)$$

which confirms the conclusion drawn above on the basis of the appearance of complex characteristics, (6.24). The higher-order dispersion term in (6.27) accounts for the first term inside the parentheses of (6.41) and thus makes the instability restricted to a finite range of sideband frequencies. The sideband growth rate is maximum for

$$\bar{\omega} = -\sigma.$$

The analysis of Benjamin and Feir (1967) thus shows that a finite-amplitude wave train may be unstable, but in order to determine the ultimate fate of the wave system, it is necessary to consider the full non-linear problem.

Soliton behavior

That finite-amplitude waves described by (6.36) may be unstable to small disturbances may at first seem surprising since such waves conserve energy. A finite-amplitude analysis shows, however, that the energy actually oscillates between different modes and that a possible end result is a wave pattern of a permanent form, a soliton.

Zakharov and Shabat (1971) have shown that (6.33) can be solved exactly by the inverse scattering method for initial conditions that approach zero sufficiently rapidly as $x \to \infty$. A terse summary of their findings was given by Yuen and Lake (1975), which we paraphrase as follows:

(a) An initial wave envelope pulse of arbitrary shape will eventually disintegrate into a number of discrete wave groups (solitons) and an oscillatory "tail." The number and structure of the solitons and the properties of the tail are completely determined by the initial conditions.

(b) The tail disperses linearly with an amplitude decay proportional to $t^{-1/2}$.

(c) Each soliton is a progressive wave solution of the form

$$A_n = a_n \operatorname{sech}\left[a_n X \sqrt{2\left(\frac{f_2}{-\Omega_{kk}}\right)}\right]$$

$$\times \exp\left\{i\left[f_2 a_n^2 + \frac{1}{\Omega_{kk}} V_n(X + \theta_n)\right]\right\}. \tag{6.43}$$

Here

$$X = x - c_{gn} t - X_n$$

and a_n and V_n characterize amplitude and speed relative to an observer moving with group velocity c_{gn} of the nth soliton. The terms X_n and θ_n give their position and phase.

(d) The solitons are stable in the sense that they can survive collisions with other solitons without any change but for a displacement in time and space.

(e) The time scale of the soliton formation is proportional to the duration of the initial disturbance and inversely proportional to the initial wave steepness.

(f) For an initial pulse in which the perturbations in ω and k are small, an estimate of the number of solitons that will form is n_s, given by the formula

$$n_s = \frac{1}{\pi} \int \sqrt{\frac{-f_2}{\Omega_{kk}}} \, a_0 f(x) \, dx, \tag{6.44}$$

where $a_0 f(x)$ is the initial profile, with $0 \le f(x) \le 1$.

One should further note that the soliton solution (6.44) has a slow time recurrence with a frequency $f_2 a_n^2$. This appears to be an analytical example of Fermi–Pasta–Ulam (1962) recurrence.

It is particularly interesting that the solutions of the nonlinear Schrödinger equation has quantum properties.

For waves propagating in two spatial dimensions, Davey and Stewartson have found the additional term for the modulation equation, and Hui and Hamilton (1979) have presented solutions.

Yuen and Lake (1980) have discussed the stability of the solutions and otherwise reviewed the subject of nonlinear modulations.

Surface gravity waves

To exemplify the application of kinematic wave theory and nonlinear modulational dynamics, we have chosen surface gravity waves. This is a wave system discussed extensively in the literature.

Consider surface gravity waves in an incompressible and inviscid fluid. If initially irrotational, the fluid will remain so, and the velocity field is derivable from a potential, with the velocity components given by

$$u_j = \frac{\partial \phi}{\partial x_j}. \tag{6.45}$$

Incompressibility requires that

$$\frac{\partial u_j}{\partial x_j} = \nabla^2 \phi = 0. \tag{6.46}$$

For sinusoidal waves propagating in the positive x direction the potential is

$$\phi = b e^{kz} \sin(kx - \omega t). \tag{6.47}$$

Here $x = x_1$ and $z = x_3$, and z is positive upward, while b is a constant.

The boundary conditions at the free surface contain the nonlinearities in the problem that cause the difficulties encountered when one attempts to proceed beyond the small-amplitude linear approximation.

Let the free surface be located at

$$z = \zeta(x, t). \tag{6.48}$$

The kinematic boundary condition may be written,

$$\frac{\partial \zeta}{\partial t} - \frac{\partial \phi}{\partial z} - \frac{\partial \phi}{\partial x} \frac{\partial \zeta}{\partial x} = 0 \quad \text{at } z = \zeta \tag{6.49}$$

(see Lamb 1932, Art. 9).

The dynamic boundary condition requires the pressure to be constant at the free surface, which amounts to

$$\frac{\partial \phi}{\partial t} + g\zeta + \tfrac{1}{2}(\nabla \phi)^2 = \text{const.} \quad \text{at } z = \zeta. \tag{6.50}$$

Note that the boundary conditions are to be satisfied at the moving free surface rather than at a fixed value of z.

To obtain a boundary condition stated at $z = 0$, we shall expand in Taylor series.

Expand (6.49) about $z = 0$ to obtain

$$\zeta_t - \phi_z - \phi_x \zeta_x - \phi_{zz} \zeta - \phi_{zzz} \tfrac{1}{2}\zeta^2 - \phi_{xz} \zeta \zeta_x + \cdots = 0 \quad \text{at } z = 0, \tag{6.51}$$

all derivatives of ϕ evaluated at $z = 0$. Similarly, a Taylor expansion of (6.50) gives

$$\phi_t + g\zeta + \tfrac{1}{2}(\phi_x^2 + \phi_z^2) + \phi_{tz}\zeta + \phi_{tzz}\tfrac{1}{2}\zeta^2 + \phi_{xz}\phi_x\zeta + \phi_{zz}\phi_z\zeta + \cdots$$
$$= \text{const.} \quad \text{at } z = 0. \tag{6.52}$$

For linear waves we drop the terms containing higher powers of ϕ and ζ in (6.51) to obtain the linear approximation of the kinematic boundary condition

$$\zeta_t = \phi_z \quad \text{at } z = 0. \tag{6.53}$$

Similarly, the linear approximation of the dynamic boundary condition (6.51) gives

$$\phi_t = -g\zeta + \text{const.} \quad \text{at } z = 0. \tag{6.54}$$

Cross-differentiation gives

$$\phi_{tt} = -g\phi_z. \tag{6.55}$$

Substitution for ϕ then yields the linear dispersion relation

$$\omega = \Omega(k) = \sqrt{gk}. \tag{6.56}$$

The waves are dispersive, with the phase speed $c = \omega/k$ dependent on the wave number k according to

$$c^2 = \frac{\omega^2}{k^2} = \frac{g}{k}.$$ (6.57)

The group velocity c_g is

$$c_g = \Omega_k = \sqrt{\frac{g}{2k}} = \frac{c}{2}.$$ (6.58)

For nonlinear waves Stokes (1847) found a periodic solution for free surface waves in deep water. He showed that the velocity potential used earlier gives a solution that also satisfies the nonlinear boundary conditions. Let

$$\Theta = kx - \omega\tau$$

denote the phase, and write the velocity potential in (6.47) as

$$\phi(x, z, t) = bce^{kz} \sin\theta,$$ (6.59)

where b is a constant. Now solve for $z = \zeta$ for the surface position from the Taylor expansion for the kinematic boundary condition (6.49). Set

$$\zeta = \zeta_1 + \zeta_2 + \zeta_3 + \cdots$$ (6.60)

to represent the successive stages in the iteration process. For $z = 0$ one finds

$$\zeta_{1t} = \phi_z,$$ (6.61)

$$\zeta_{2t} = \phi_x \zeta_1 + \phi_{zz} \zeta_1,$$ (6.62)

$$\zeta_{3t} = \phi_x \zeta_2 + \phi_{zz} \zeta_2 + \tfrac{1}{2}\phi_{zzz} \zeta_1^2 + \phi_{xz} \zeta_1 \zeta_{1x},$$ (6.63)

and so on.

After extensive algebra one obtains (see Lamb 1932, Art. 250)

$$\zeta = \tfrac{1}{2}kb^2 + b(1 + \tfrac{9}{8}kb)\cos\Theta + \tfrac{1}{2}kb^2\cos 2\Theta + \tfrac{3}{8}k^2b^3\cos 3\Theta + \cdots$$ (6.64)

Now set

$$a = b(1 + \tfrac{9}{8}kb),$$ (6.65)

so that (6.64) may be written as

$$\zeta = \tfrac{1}{2}a^2k + a\cos\Theta + \tfrac{1}{2}a^2k\cos 2\Theta + \tfrac{3}{8}a^3k^2\cos 3\Theta + \cdots.$$ (6.66)

Note that this is a Fourier series in phase Θ, and that the phase propagates at constant speed, so that the wave keeps its shape as it propagates. The waves are thus waves of "permanent form."

Next apply the dynamic boundary condition (6.52) using (6.51). After some algebra one finds

$$ck(1-k^2b^2)-g=O(k^3b^3). \qquad (6.67)$$

The phase speed c to order $O(\epsilon^2)$ ($\epsilon=ka$) is, to the same order of approximation,

$$c=\frac{g}{k}\left(1+\frac{k^2a^2}{2}\right). \qquad (6.68)$$

The frequency is

$$\omega=[gk(1+\tfrac{1}{2}k^2a^2)]^{1/2}. \qquad (6.69)$$

Hence, for the nonlinear contribution, f_2 [see (6.23)], to the dispersion relation, one finds

$$f_2=k^{5/2}g^{1/2}. \qquad (6.70)$$

Also, from the linear dispersion relation it follows that

$$\Omega^r_{kk}=-\tfrac{1}{4}g^{1/2}k^{-3/2}. \qquad (6.71)$$

Thus,

$$f_2\Omega^r_{kk}<0, \qquad (6.72)$$

and according to the Benjamin–Feir (1967) instability theory, a finite-amplitude surface gravity wave on an infinitely deep water is hence susceptible to sideband instability. For finite-depth waves Benjamin and Feir found instability only for $kH>1.3$.

Discussion

The analysis presented shows that weak nonlinearities can, given time and distance to develop, exercise a dominant influence on the wave field. This also shows that the effects of nonlinearity can only be observed over long distances and times, so that a laboratory experiment, for example, has to be carefully designed before one can observe clear nonlinear effects. For gravity waves on the open sea, however, there is plenty of time and distance available for nonlinear development, and one should expect such waves to have fully developed nonlinear modulational characteristics. If one observes waves over limited times and distances, one may well conclude that linear theory is adequate as a description of one's observations; this does not prove that the nonlinear effects are negligible under all circumstances.

From (6.31) it follows that for waves propagating through a spatially varying background (through a region of variable current, for example) with $\Omega_k^r = O(\epsilon)$, nonlinearities will become important for amplitudes (wave steepnesses) of order $\epsilon^{1/2}$ or greater. Conversely, for waves of low amplitude nonlinearity will be correctly described by equations like (6.32) only when the background inhomogeneity is small compared to the square of the wave amplitude. The applicability of the Ginsburg–Landau and cubic Schrödinger equations to realistic flow situations is therefore somewhat limited. Nevertheless, it is significant that the Ginsburg–Landau equation can be demonstrated to show chaotic behavior (Moon et al. 1983). Also, Hocking, Stewartson, and Stuart (1972) demonstrated that it can give solutions of infinite amplitudes within a finite time.

The more complete kinematic wave theory also suggests that other small influences can have profound modulational consequences. Bliven, Huang, and Long (1985) found that the effect of wind on a periodic wave train can eliminate the sideband instability. Now, if one considers the atmospheric wake of a wave group with flow separation over the crests, one should also suppose that such a periodic wake should affect the stability of wave modulations further downwind. The atmospheric boundary layer has other instabilities and can support disturbances with a dispersion different from that of water waves. When waves break, their momentum goes into a current, which in turn affects the modulational behavior of the waves that follow. With the multiplicity of possible couplings between disturbances and wave modulations, one can understand how a wind wave field can rapidly become chaotic. As a consequence, one cannot expect that it will be possible to develop deterministic wave prediction models valid over large times and distances. Stochastic models may be possible, but even these may have to include effects of time history of the wave field and the effects of preexisting waves on local wave development. Models local in time and space may still be successful in a sense, but it seems to us that reliance on empirical models may be necessary. For the very strongest winds, it is possible and even likely that the wave field will be totally dominated by the local wind, so that the most intense part of the wave field may be predictable.

The basis for the nonlinear modulation theory presented here has been established in several experimental investigations. Huang et al. (1981, 1983, 1984) have shown that mechanically generated water waves and wind waves are indeed Stokes waves. This confirms the finding of Ramamonjiarisoa (1974) that wind waves have a dispersion relation that suggests that they are waves of permanent form. Su et al. (1982) showed that waves generated by an oscillating plunger would form groups, and Chereskin and Mollo-Christensen (1985) demonstrated that their modulational dynamics could be described in terms of a nonlinear Schrödinger equation with an additional damping term.

The representation of random wave fields in terms of independent random solitons was suggested by Mollo-Christensen and Ramamonjiarisoa (1978, 1982) who obtained results that in some respects were reminiscent of observations of oceanic waves.

In addition to the modulational instabilities described here, there are other wave instabilities. Longuet-Higgins (1978) found a subharmonic instability to be possible for large wave steepness. Melville (1983) found a similar process experimentally. There are still a number of nonlinear wave phenomena that have not been mentioned, and the field is under continuous development. Of special interest is the behavior of random nonlinear fields, which would apply to turbulence, wind waves, weather patterns, and oceanic motions.

The predictability of such fields and measures of fields of these kinds are likely to continue to be subjects of great interest and importance for some time to come.

It should be emphasized that the kinematic wave theory and its extensions only cover some limited aspects of wave propagation, such as interaction with a nonuniform background and effects of weak dissipation and nonlinearity on the propagation characteristics. There are many other interesting aspects of wave motion, such as (resonant or nonresonant) interaction between waves of different wave numbers and nonlinear distortion. For such problems no such general approach is available, but one needs instead to treat each wave system separately.

7

Instability and transition to turbulence

Flows of fluids of low viscosity may become unstable when large gradients of kinetic and/or potential energy are present. The flow field set up by the instability generally tends to smooth out the velocity and temperature differences causing it. The available kinetic or potential energy released by the instability may be so large that transition to a fully developed turbulent flow occurs.

Transition is influenced by many parameters. An important one is the level of preexisting disturbances in the fluid; a high level would generally cause early transition. Another cause for early transition in the case of wall-bounded shear flows is surface roughness. The manner in which transition occurs may also be very sensitive to the detailed flow properties.

For shear flows the basic nondimensional flow parameter measuring the tendency toward instability and transition is the Reynolds number; for high Re values, kinetic energy differences can be released faster into fluctuating motion than viscous diffusion will have time to smooth them out. For a heated fluid subject to gravity the Rayleigh number is the main stability parameter.

Of crucial importance for the tendency of a flow to become unstable and go through transition is the detailed distribution of mean velocity and/or temperature in the field. The analysis that follows is intended to illustrate this.

Although the flow processes involved in instability and transition might at a first glance appear to have only a slight resemblance to those observed in fully developed turbulence, they are nevertheless related to it in important ways. In a gross sense turbulence may be regarded as a manifestation of flow instability occurring randomly in space and time. The linear instability problem is the simplest flow model incorporating the interaction between unsteady fluctuations and a background shear or density distribution. With the aid of nonlinear instability theory one may also possibly be able to clarify some of the mechanisms whereby turbulence is maintained.

80

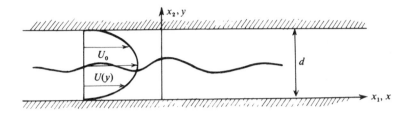

Figure 7.1. Instability of two-dimensional channel flow.

Instability to small disturbances

Because of the mathematical difficulties in the analysis of flow instability, only idealized cases for which the basic fluid flow properties vary with one spatial coordinate can be analyzed in a reasonably simple manner. Such cases include parallel shear flows, which may be used as approximate models for boundary layers and free shear layers, and a heated uniform horizontal layer of infinite extent, the simplest model for the onset of thermal convection.

Instability to infinitesimal disturbances, a problem that may be treated with the aid of linearized theory, is but an initial step in a complicated series of events leading to transition to fully developed turbulence. In fact, linear stability theory has so far been of limited value for predicting transition. Nevertheless, stability theories give important insight into mechanisms whereby fluctuations may arise in a flow with time-independent properties.

Linear instability of a parallel shear flow

Consider small (infinitesimal) perturbations in a parallel shear flow as exemplified by the flow in a two-dimensional channel (Figure 7.1). We assume the flow to consist of a mean parallel flow plus perturbations represented by

$$U_i = \bar{U}_i + u_i, \qquad P = \bar{P} + p, \tag{7.1}$$

where the mean flow is given by

$$\bar{U}_i = U(x_2)\delta_{1i}. \tag{7.2}$$

Equations governing the perturbation quantities u_i and p are those given by (3.25). We take the fluid to be of constant density (the instability due to density perturbations caused by temperature variations will be treated separately). For infinitesimal disturbances we may neglect terms that are quadratic in the perturbation quantities. The linearized equations governing the perturbations then become

$$\frac{\partial u_i}{\partial t} + U\frac{\partial u_i}{\partial x} + vU'\delta_{1i} = -\rho^{-1}\frac{\partial p}{\partial x_i} + v\frac{\partial^2 u_i}{\partial x_j\,\partial x_j}, \tag{7.3}$$

$$\frac{\partial u_i}{\partial x_i} = 0, \tag{7.4}$$

where $x = x_1$, $v = u_2$, and $U' = dU/dy$. We first take $\partial/\partial x_i$ (the divergence) of (7.3). With the aid of the continuity equation (7.4) this yields

$$\frac{\partial^2 p}{\partial x_i\,\partial x_i} = \nabla^2 p = -2\rho U'\frac{\partial v}{\partial x}. \tag{7.5}$$

This may be used to eliminate the pressure from the second component $(i = 2)$ or (7.3),

$$\frac{\partial v}{\partial t} + U\frac{\partial v}{\partial x} = -\left(\frac{1}{\rho}\right)\frac{\partial p}{\partial y} + v\nabla^2 v. \tag{7.6}$$

Taking ∇^2 of this and combining it with $\partial/\partial y$ of (7.5), we find, after some simplifications,

$$\left(\frac{\partial}{\partial t} + U\frac{\partial}{\partial x}\right)\nabla^2 v - U''\frac{\partial v}{\partial x} - v\nabla^4 v = 0. \tag{7.7}$$

The boundary conditions for the problem are

$$u, v, w = 0 \quad \text{on } y = 0, d. \tag{7.8}$$

Equations (7.7) and (7.8) constitute a homogeneous boundary value problem that, in principle, could be solved for any arbitrary initial condition

$$v(x, y, z, 0) = v_0(x, y, z), \tag{7.9}$$

satisfying the boundary conditions (7.8). Thereby, one could determine whether an arbitrary initial infinitesimal velocity disturbance will grow indefinitely. Since in general this is quite a complicated problem, we shall instead make use of the homogeneity of the system (7.7) and (7.8) in x, z, t and seek separable solutions in the form of infinite traveling wave trains (or Fourier components) of the form

$$v(x, y, z, t) = \hat{v}(y)\exp[i(\alpha x + \beta z - \omega t)], \tag{7.10}$$

where α and β are the streamwise and spanwise wave numbers, respectively, and ω is the frequency. We will also set

$$\omega = \alpha c, \tag{7.11}$$

where c is (the streamwise component of) the phase velocity. Substitution of (7.10) into (7.7) yields

$$(U-c)\left(\frac{d^2\hat{v}}{dy^2} - k^2\hat{v}\right) - U''\hat{v} - \frac{\nu}{i\alpha}\left(\frac{d^2}{dy^2} - k^2\right)^2\hat{v} = 0, \tag{7.12}$$

where $k^2 = \alpha^2 + \beta^2$.

At this stage it is convenient to introduce nondimensional variables by dividing all velocities by a suitable reference velocity, for example, by the maximum velocity U_0 (free-stream velocity in the case of a boundary layer). Also, all spatial variables are divided by a reference length, for example, d. This will simply lead to the replacement of ν in (7.12) by its nondimensional equivalent, $1/R$, where the Reynolds number R is defined by

$$R = \frac{U_0 d}{\nu}. \tag{7.13}$$

One thus finds for $\hat{v}$ the following problem:

$$(U-c)\left(\frac{d^2\hat{v}}{dy^2} - k^2\hat{v}\right) - \hat{v}U'' - \frac{1}{i\alpha R}\left(\frac{d^2}{dy^2} - k^2\right)^2\hat{v} = 0, \tag{7.14}$$

$$\hat{v}(0) = \hat{v}'(0) = \hat{v}(1) = \hat{v}'(1) = 0. \tag{7.15}$$

When formulating the problem for a boundary layer, a suitable length scale is the boundary layer thickness δ or the displacement thickness δ_*. Furthermore, the second set of boundary conditions are to be applied at $y = \infty$ rather than at $y = 1$.

Equations (7.14) and (7.15) define a homogeneous boundary value problem that can only have nontrivial solutions $\hat{v} \neq 0$ for certain characteristic values of the parameters. We shall assume that α, β, and Re are given (real) values and that c is the sought characteristic value. Generally, there would be more than one such characteristic value,

$$c = c^{(n)} = c_r^{(n)} + ic_i^{(n)}. \tag{7.16}$$

The wavelike disturbance is unstable if for any of the eigenvalues $c_i^{(n)} > 0$ because

$$\exp(-i\alpha ct) = \exp(-i\alpha c_r t)\exp(\alpha c_i t). \tag{7.17}$$

The problem thus presented is the famous Orr–Sommerfeld stability problem (Orr 1907; Sommerfeld 1908), which has received a great deal of attention over the years, as mentioned in Chapter 1. With the use of

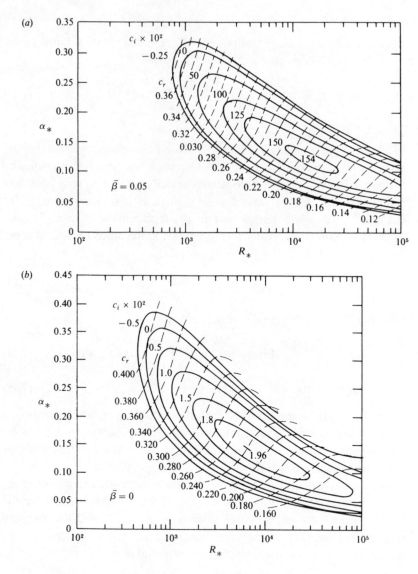

Figure 7.2. (*a*) and (*b*).

modern computing techniques accurate solutions may be generated for a large number of cases. Such examples are shown in Figure 7.2 (taken from Obremski, Morkovin, and Landahl 1969) computed for Falkner–Skan self-similar boundary layer profiles having a free-stream velocity varying in the streamwise direction as

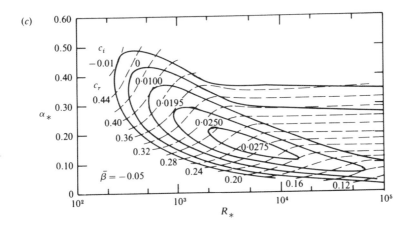

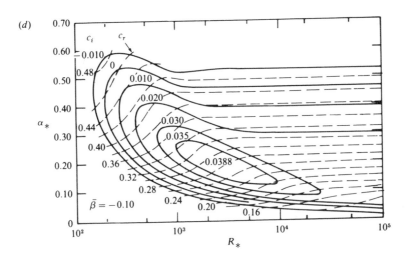

Figure 7.2. Curves of constant temporal amplification rates for Falkner–Skan velocity profiles: (*a*) $\bar{\beta} = 0.05$; (*b*) $\bar{\beta} = 0$; (*c*) $\bar{\beta} = -0.05$; (*d*) $\bar{\beta} = -0.10$ (from Obremski et al. 1969).

$$U_0 \sim x^m, \tag{7.18}$$

where $m = \bar{\beta}/(2 - \bar{\beta})$, $\bar{\beta}$ being a pressure gradient parameter that is negative for flows with positive ("adverse") pressure gradient and positive for flows with negative ("favorable") pressure gradient. The Blasius flat-plate boundary layer is the special case of $\bar{\beta} = 0$. The results shown are made dimensionless through the use of the displacement thickness

$$\delta_* = \int_0^\infty \left(1 - \frac{U}{U_0}\right) dy \tag{7.19}$$

as a reference length. Thus

$$R_* = \frac{U_0 \delta_*}{\nu}, \qquad \alpha_* = \alpha \delta_*. \tag{7.20}$$

Of course, for a boundary layer the Orr–Sommerfeld theory holds only approximately since the flow in a boundary layer is not strictly parallel. However, the streamwise variation is so slow that as a good first approximation one may neglect its influence on stability.

It is seen from the diagrams (only the lowest order = least stable eigenvalues are shown) that the flows with $\bar{\beta} < 0$ are considerably more unstable than those with $\bar{\beta} > 0$. Also, the values of c_r and c_i tend to become independent of R_* for large R_* when $\bar{\beta} < 0$. This has to do with the appearance of inflection in the velocity profiles for boundary layers in a positive pressure gradient. The importance of inflection was realized already by Rayleigh in his (1878) study of the stability of an inviscid flow.

For the inviscid case $\nu = 0$ (Re $= \infty$) we have

$$(U - c)\left(\frac{d^2 \hat{v}}{dy^2} - k^2 \hat{v}\right) - U'' \hat{v} = 0 \tag{7.21}$$

with the boundary conditions

$$\hat{v}(0) = \hat{v}(1) = 0. \tag{7.22}$$

Dividing the equation by $U - c$, multiplying it with the complex conjugate $\hat{v}^*$, and integrating it between $y = 0$ and 1, we find, after an integration by parts and making use of the boundary conditions (7.22),

$$\int_0^1 \left(\left|\frac{d\hat{v}}{dy}\right|^2 + k^2 |\hat{v}|^2\right) dy + \int_0^1 \frac{U'' |\hat{v}|^2}{U - c} \, dy = 0. \tag{7.23}$$

Taking the imaginary part of this, one finds

$$c_i \int_0^1 \frac{U'' |\hat{v}|^2}{(U - c_r)^2 + c_i^2} \, dy = 0, \tag{7.24}$$

which for $c_i = 0$ can only be satisfied if U'' changes sign in the interval, that is, if the velocity distribution possesses a point of inflection. A later investigation by Tollmien (1929, 1935) showed that the necessary condition of $U'' = 0$ is also a sufficient condition for instability for a large class of flows. Fjørtoft (1950) also demonstrated that for the flow to be unstable, $U'' = 0$ must correspond to a point of maximum shear. From such general results one may surmise that whenever there is an inflection point in the velocity distribution of a high-Reynolds-number shear flow, instability can be expected to occur. The instability becomes particularly strong for free-shear layers and separated boundary layers. One can make use of such information for the design of low-drag airfoils; so-called laminar flow airfoils are shaped such that they have their maximum thickness far back in order to achieve as long a region of favorable (i.e., negative) pressure gradient as possible (cf. also the shapes of fast-swimming or fast-flying animals).

The experimental verification of the stability theory took very long. It was first Schubauer and Skramstad (1947) who, by exciting waves of a given frequency with the aid of a vibrating ribbon, were able to show that the experiment gave stability characteristics in good agreement with the theory.

Later experiments by other investigators using the vibrating ribbon technique have given additional confirmations of the basic tenets of the theory. In Figures 7.3 and 7.4 are reproduced the stability boundary and amplification rates obtained by Ross et al. (1970). As seen, the agreement between theory and experiments is quite good for the amplification rates; for the stability boundary there is some discrepancy between theory and experiments in the region near the critical Reynolds number (the lowest Reynolds number for which there is instability).

The numerical results shown are for two-dimensional waves only, that is, waves for which $\beta = 0$. As pointed out by Squire (1933), only two-dimensional waves need be considered, since the solution of the Orr–Sommerfeld problem for a wave with $\beta \neq 0$ may be obtained from that for $\beta = 0$ by solving the problem for $\alpha = k$ and $R_1 = R\alpha/k$ (Squire's transformation). By applying this to the Blasius boundary layer, Squire (1933) found that two-dimensional waves actually were the first to go unstable, so that the critical Reynolds number for instability obtained from two-dimensional waves also is valid for three-dimensional ones.

By a slight extension of Squire's transformation one could also analyze the stability of three-dimensional parallel flows of the form $[U(y), 0, W(y)]$. Following the same procedure as in the derivation of the Orr–Sommerfeld equation, one then finds that the stability to small disturbances is governed by the equation

$$(\alpha U + \beta W - \omega)\left(\frac{d^2\hat{v}}{dy^2} - k^2\hat{v}\right) - (\alpha U'' + \beta W'')\hat{v}$$

$$- \frac{1}{iR}\left(\frac{d^2}{dy^2} - k^2\right)^2 \hat{v} = 0. \qquad (7.25)$$

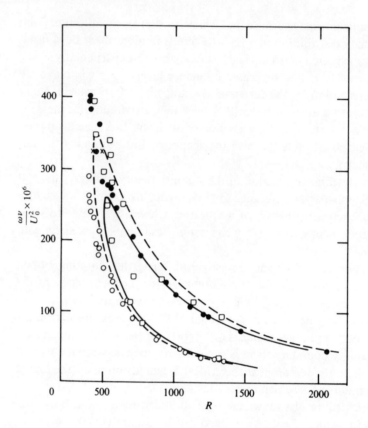

Figure 7.3. Comparison of experiments and theories for neutral stability curves for a Blasius boundary layer. - - -, theory by Shen (1954); ——, theory by Ross et al. (1970) (from Ross et al. 1970).

It is easily seen that this may be written as the ordinary Orr–Sommerfeld equation (7.12) by the introduction of the transformed velocity $U_1 = U + \beta W/\alpha$ and all the general results for that equation made use of, including the inflection point criterion for inviscid stability. Thus, one finds that the boundary layer of a swept wing may be susceptible to cross-flow instability because of inflection in the W-velocity profile.

Linear instability of a layer heated from below

Thermal (convective) turbulence has its origin in the instability due to density differences caused by differential heating. Because of the restraining effect of internal friction and the slow, but finite, thermal diffusion

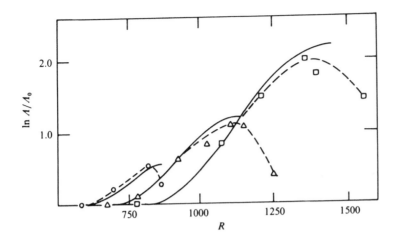

Figure 7.4. Amplification rates for wave trains of constant frequency (from Ross et al. 1970).

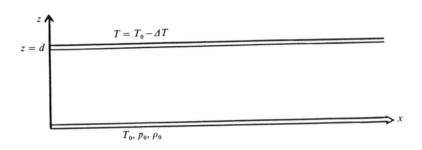

Figure 7.5. Layer of fluid heated from below, definitions.

rate, a viscous and heat-conducting fluid can sustain a certain amount of unstable density stratification, i.e., with the heavier fluid on top, before an overturning motion sets in. The simplest model problem illustrating this behavior is that of a Boussinesq fluid between two horizontal infinite parallel plates, the lower one of which is held at a slightly higher temperature than the upper one (Figure 7.5).

Following the usual convention in the literature on this subject, we shall take the z axis to be in the vertical. The lower and upper plates are held at fixed temperatures T_0 and $T_0 - \Delta T$, respectively, and the pressure and density of the fluid at the lower plate is taken to be p_0 and ρ_0, respectively. The Boussinesq model equations (3.13), (3.7), (3.11) for this case become

$$\rho_0\left(\frac{\partial U_i}{\partial t}+U_j\frac{\partial U_i}{\partial x_j}\right)=-\frac{\partial P}{\partial x_i}-g\rho_0[1-\alpha(T-T_0)]\delta_{i3}+\mu\nabla^2U_i, \qquad (7.26)$$

$$\frac{\partial U_i}{\partial x_i}=0 \qquad (7.27)$$

$$\frac{\partial T}{\partial t}+U_j\frac{\partial T}{\partial x_j}=\kappa_h\nabla^2T. \qquad (7.28)$$

The boundary conditions are

$$T=T_0, \qquad U_i=0, \quad \text{for } z=0, \qquad (7.29)$$

$$T=T_0-\Delta T, \qquad U_i=0, \quad \text{for } z=d. \qquad (7.30)$$

We now introduce

$$P=\bar{P}+p, \qquad (7.31)$$

$$U_i=u_i, \qquad (7.32)$$

$$T=\bar{T}+\theta, \qquad (7.33)$$

which yield the following solutions for the mean temperature, pressure and density:

$$\bar{T}=T_0-\frac{\Delta Tz}{d}=T_0-\beta z, \qquad (7.34)$$

$$\bar{P}=p_0-g\rho_0(z+\tfrac{1}{2}\alpha\beta z^2), \qquad (7.35)$$

$$\bar{\rho}=\rho_0(1+\alpha\beta z). \qquad (7.36)$$

Equations for the disturbance quantities are obtained by subtracting the equations for the mean from the full equations. Nonlinear terms are neglected on the assumption that the fluctuations are infinitesimal in amplitude. This yields

$$\frac{\partial u_i}{\partial t}=-\frac{\rho_0^{-1}\partial p}{\partial x_i}+g\alpha\theta\delta_{i3}+\nu\nabla^2u_i, \qquad (7.37)$$

$$\frac{\partial u_i}{\partial x_i}=0, \qquad (7.38)$$

$$\frac{\partial\theta}{\partial t}=\beta w+\kappa_h\nabla^2\theta. \qquad (7.39)$$

Also, by taking the divergence $(\partial/\partial x_i)$ of (7.37), we find

$$\nabla^2p=\rho_0 g\alpha\frac{\partial\theta}{\partial z}. \qquad (7.40)$$

Next, we employ (7.40) to eliminate the pressure from the third of (7.37). This gives

$$\frac{\partial}{\partial t}\nabla^2 w = g\alpha\left(\frac{\partial^2\theta}{\partial x^2}+\frac{\partial^2\theta}{\partial y^2}\right)+\nu\nabla^4 w. \tag{7.41}$$

We may use (7.39) to express w in terms of θ, which then yields a single equation for θ,

$$\nabla^2\left(\nu\nabla^2-\frac{\partial}{\partial t}\right)\left(\frac{\partial\theta}{\partial t}-\kappa_h\nabla^2\theta\right)+g\alpha\beta\left(\frac{\partial^2\theta}{\partial x^2}+\frac{\partial^2\theta}{\partial y^2}\right)=0. \tag{7.42}$$

The boundary conditions are that

$$\theta=9, \qquad u,v,w=0, \quad \text{at } z=0,d. \tag{7.43}$$

From (7.39) we then find

$$\frac{\partial\theta}{\partial t}-\kappa_h\nabla^2\theta=0 \quad \text{at } z=0,d. \tag{7.44}$$

Differentiating (7.39) with respect to z and using continuity, according to which

$$\frac{\partial w}{\partial z}=-\frac{\partial u}{\partial x}-\frac{\partial v}{\partial y}=0 \quad \text{at } z=0,d, \tag{7.45}$$

we obtain the additional condition

$$\left(\frac{\partial}{\partial t}-\kappa_h\nabla^2\right)\frac{\partial\theta}{\partial z}=0 \quad \text{at } z=0,d. \tag{7.46}$$

From (7.42), (7.44), and (7.46) we have the necessary six boundary conditions for the temperature fluctuations θ that (together with suitable initial conditions) specify the problem. In a manner similar to that used in the shear flow problem, we look for solutions of the form

$$\theta=\hat\theta(z)\exp[\lambda t+i(k_1 x+k_2 y)], \tag{7.47}$$

which, when inserted in (7.41), (7.42), (7.44), and (7.46), give

$$(D^2-k^2)[\nu(D^2-k^2)-\lambda][\lambda\hat\theta-\kappa_h(D^2-k^2)\hat\theta]-g\alpha\beta k^2\hat\theta=0, \tag{7.48}$$

with boundary conditions

$$\hat\theta=[\lambda-\kappa_h(D^2-k^2)]\hat\theta=[\lambda-\kappa_h(D^2-k^2)]D\hat\theta=0 \quad \text{for } z=0,d, \tag{7.49}$$

and where $D=d/dz$.

The stability boundary is reached as the real part of λ goes to zero. One can show that the eigenvalues $\lambda^{(n)}$ are all real, so that it is only necessary to consider $\lambda=0$ in (7.48) and (7.49). This leads to the simplified problem

$$(D_*^2 - a^2)^3 \hat{\theta} + a^2 \, \mathrm{Ra} \, \hat{\theta} = 0, \qquad (7.50)$$

$$\theta = D_*^2 \hat{\theta} = (D_*^2 - a^2) D_* \hat{\theta} = 0, \qquad (7.51)$$

where the equation has been non-dimensionalized by introduction of $z^* = z/d$, $D_* = d/d_* z$, and where

$$a^2 = k^2 d^2, \qquad (7.52)$$

$$\mathrm{Ra} = \frac{g \alpha \beta d^4}{\nu \kappa_h} = \frac{g \alpha \, \Delta T \, d^3}{\nu \kappa_h} \quad \text{(the Rayleigh number)}. \qquad (7.53)$$

(We will omit the asterisk in the following.) The solutions of the homogeneous equation (7.50) are of the form

$$\theta_m = A_m \exp(b_m z), \qquad m = 0, 1, \ldots, 5, \qquad (7.54)$$

where b_m are the six roots of the equation

$$(b^2 - a^2)^3 + a^2 \, \mathrm{Ra} = 0, \qquad (7.55)$$

that is,

$$b_m = (a^2 - a^{2/3} \, \mathrm{Ra}^{1/3} \, e^{2\pi i m/3})^{1/2} (-1)^m. \qquad (7.56)$$

By introduction of these into the boundary conditions (7.51), one obtains an eigenvalue relation for the wave number a as a function of the Rayleigh number, Ra. One can distinguish between even (with respect to the midpoint $z_* = \frac{1}{2}$) and odd eigensolutions, the even ones yielding the lowest critical Rayleigh number of $\mathrm{Ra}_{cr} = 1708$ for $a = 3.12$ (see Figure 7.6). Experiments have given excellent agreement with this value for the onset of convection.

An extensive treatment of the problem of convective instability ("Bénard problem") may be found in the book by Chandrasekhar (1961).

Universality in transition to chaos

Instability to small disturbances is but the first stage in a complicated sequence of events leading to transition to a fully developed turbulent flow, that is, one that contains the full range of eddy scales with random time and space dependence. The subsequent nonlinear stage of transition may proceed in many different ways. Not only do different flows follow different paths to turbulence, but one and the same flow may vary in its transition behavior depending, for example, on the level and nature of preexisting disturbances in the flow and on how the transition is initiated.

Random, "chaotic", behavior is not unique to turbulent flows. Such behavior has also been observed in mechanical, electrical, chemical, and

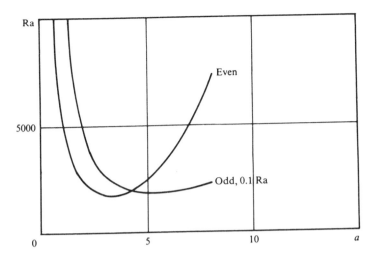

Figure 7.6. Stability boundary of a heated fluid layer between two parallel and perfectly conducting plates (after Chandrasekhar 1961).

biological systems, most of which are considerably less complex than flows of a continuum fluid. In fact, chaotic behavior appears to be very common in nonlinear systems. The manner in which relatively simple nonlinear systems become chaotic is a subject that has been much studied. For a recent review of this subject, see, for example, Helleman (1980).

The chaotic behavior of systems that satisfy deterministic equations can be attributed to the mathematical structure of the equations. The equations have the property that, for certain ranges of parameters and variables, they are singularly dependent upon initial conditions. This means that an infinitesimal change in initial conditions can cause a finite change in the solution. A system of this nature with many, even an infinity of, degrees of freedom (as for a fluid) will then show an unpredictable behavior, while it satisfies all the conservation principles exactly. Chaotic continua are of course very complicated, and work on chaotic systems has, so far, been confined to studies of discrete, simple systems of a finite, usually small number of degrees of freedom. Chaotic behavior of such systems is characterized by aperiodic or quasi-periodic motion producing a continuous and fairly flat frequency spectrum. The systems studied have been found to exhibit a certain degree of universal behavior in their transition from periodic to chaotic motion. One has identified (at least) three different "scenarios" for the transition to chaos as a parameter measuring the nonlinearity of the system (such as the Rayleigh number in the case of thermal convection) is increased, namely: (i) succession of bifurcations to subharmonic instabilities until an accumulation point in the nonlinearity parameter is reached (Feigenbaum 1980); (ii) a finite number (three or four) of (Hopf-type) bifurcations before chaotic behavior sets in abruptly (Ruelle and Takens 1971); and (iii) intermittent transition (Manneville and Pomeau 1980).

In the Feigenbaum (1980) scenario the nonlinear solution bifurcates to subharmonic motion with period doubling at each passage of a critical value of the nonlinear parameter. The critical values appear ever more densely packed as the parameter is increased until an acculumlation point is reached at which the lower end of the spectrum becomes completely saturated with subharmonic oscillations. For values of the parameter beyond this point, chaos sets in with the peaks at the subharmonic frequencies eventually disappearing. Feigenbaum was able to demonstrate with the aid of a simple diagram, a "map," showing the amplitude maximum in one cycle plotted against the maximum in the previous cycle, that the subharmonic transitions follow a universal behavior that is independent of the particular nonlinear system studied. Thus, he found that the critical values R_n of the nonlinearity parameter R for bifurcation to a new subharmonic follow a geometric series that, as n tends toward infinity, has the following behavior:

$$\frac{R_{n+1}-R_n}{R_{n+2}-R_{n+1}} = \delta \qquad (n \to \infty), \tag{7.57}$$

where $\delta = 4.66920\ldots$ is Feigenbaum's first universal constant (obtained by numerical experimentation). He also discovered a second universal constant measuring the ratio of the amplitudes of successive subharmonic bifurcations as

$$\lim_{n \to \infty} \frac{A_{n+1}}{A_n} = \alpha = 2.50920\ldots, \tag{7.58}$$

where A_n is the nth amplitude maximum of the quasiperiodic motion.

A simple dynamical system found to follow Feigenbaum's scenario is that described by the Duffing equation,

$$\frac{d^2x}{dt^2} + \frac{D\,dx}{dt} + x + x^3 = b\cos(2\pi\omega t), \tag{7.59}$$

where D is a damping constant.

The Ruelle–Takens (1971) scenario is concerned with "generic" systems exhibiting *Hopf-type* bifurcations. In such systems the solution bifurcates to a limit cycle from a stable fixed point in the phase plane (see Figure 7.7(a)). This is in contrast to bifurcation to *subharmonic* motion when

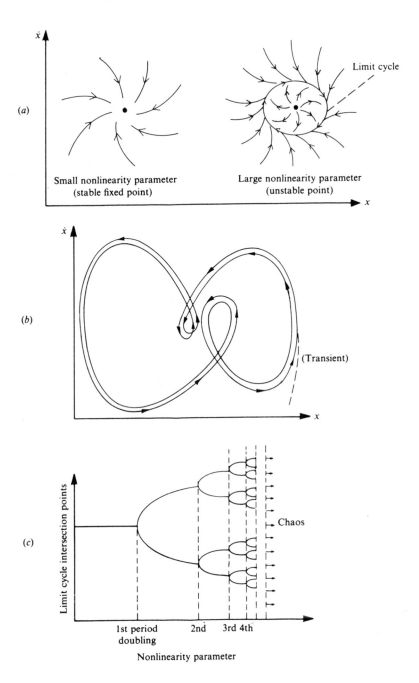

Figure 7.7. Phase portraits of bifurcations in a nonlinear oscillator. (*a*) Hopf-type bifurcation; (*b*) two-cycle oscillation; (*c*) pitchfork-type bifurcation.

a limit cycle changes over to one with a double period; in the phase plane the orbit (for a one-dimensional system) will look like a pretzel (see Figure 7.7(b)). If the maximum amplitude of the motion is plotted versus the nonlinearity parameter, as the subharmonic motion arises, the curve will look like a tree with ever more fine branching as the parameter is increased (see Figure 7.7(c)). This is a so-called *pitchfork-type* bifurcation. Ruelle and Takens (1971) found that at most four (later shown to be three) Hopf-type bifurcations are likely to occur before nonperiodic and chaotic motion sets in.

In the Manneville–Pomeau (1980) scenario an intermittent transition to chaos takes place when a stable and an unstable fixed point in the phase plane collide as the nonlinearity parameter is increased. This scenario is the least well understood and the most difficult to study since it does not show any clear-cut precursors.

Clearly, the scenarios thus described are quite different from that proposed by Landau (1944), believed earlier to give a general qualitative description of the transition process. According to this, each new instability arising as the nonlinearity parameter is increased produces an added limit cycle (the nonlinearity assumed to be amplitude limiting) with an additional higher frequency. Eventually, the spectrum, although discrete, will become so dense that it appears indistinguishable from a continuous one. None of the simple nonlinear systems that have been studied to date have clearly exhibited such a behavior, however, and Landau's scenario is now generally considered untenable.

In recent years a great deal of work has also been expended on the post-transitional behavior of simple chaotic systems in the hope that studies of such systems would shed light on turbulence in continuum systems. Lorenz (1963) modeled thermal convection in a Boussinesq fluid by a system of three coupled first-order equations with quadratic nonlinearities corresponding to the representation of the continuous system by three modes. He found that for a Rayleigh number greater than about 25 times the critical value for the first linear instability, the motion became chaotic in such a manner that it switched in an erratic manner between two almost periodic limit cycle type motions. The detailed path in the phase plane was found to be extremely sensitive to the initial conditions. This has become known as a *strange attractor* behavior. An ordinary attractor in the phase plane is either a fixed point (stable, steady motion) or a closed curve (limit cycle). A repellor may be a fixed point or a closed curve from which the motion spirals outward with increasing amplitude. A strange attractor possesses both attracting and repelling behavior. In the case

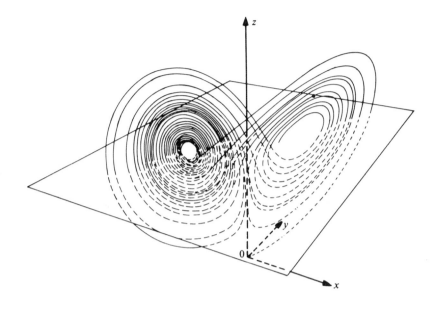

Figure 7.8. Poincaré map for the Lorenz (1963) oscillator showing strange attractor behavior.

of a strange attractor a *Poincaré map,* that is, a lower-dimensional cut through the phase space, will appear as a cloud of points, or lines, showing the intersection of the motion paths with the cutting plane. In an aperiodic motion subsequent paths will never intersect the plane in the same point as a preceding one, although the points may come infinitesimally close (see Figure 7.8).

How the behavior of simple nonlinear systems showing chaotic behavior can shed light on the problem of transition to turbulence in flows of continuum fluids is presently a debated issue. What is most typical of transition in fluid flows is the appearance of the small scales of motion, which for shear flows is generally a fairly abrupt process. To represent the transitional behavior of such flows by that of simple systems with a small number of degrees of freedom is therefore not appropriate. For thermal convection, and for Taylor–Couette flow between rotating cylinders, on the other hand, the transition is a more gradual one with initially a relatively simple spatial structure that may be described by a small number of modes. The temporal behavior of such flows might therefore be reasonably well described by the simple nonlinear systems showing chaotic behavior as described above, as has indeed been verified by experiments with

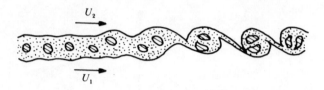

Figure 7.9. Vortex pairing in a free-shear layer (conceptual).

such flows. In the next section we shall review observations of transitions in different flows and point out possible connections to the alternative scenarios described above. For a more complete discussion of this and other related questions, see Monin (1978).

Transition in a free-shear layer

Because of its inflectional velocity profile, a free-shear layer is unstable to small disturbances already at very low Reynolds numbers. In such a layer instability waves will therefore readily appear. Although, according to linear theory, there is a band of wave numbers that are unstable, a wave number near the most amplified one is usually found to be dominating, producing a row of vortices that in flow visualization experiments appear to be more or less rolled up (Figures 7.9 and 7.10). In the finite-amplitude development that ensues, two different nonlinear processes have been found to take place. First, there is a growth of subharmonics [waves of a wave number half of that of the primary instability wave (Kelly 1967)], appearing in visualization experiments as pairing of neighboring vortices (Winant and Browand 1974), as illustrated in Figure 7.9. Secondly, the finite-amplitude row of vortices develops longitudinal streaks (Bernal 1981), which may be the manifestation of a secondary instability of oblique waves developing on the two-dimensional periodic flow (Pierrehumbert and Widnall 1982). Within these fairly large scale vortical disturbances small-scale turbulence develops, leading eventually to a fully developed turbulent shear flow.

Transition in a boundary layer

How a boundary layer undergoes transition has been quite extensively studied during the last 50 years or so, but many of the underlying mechanisms are still incompletely understood. We shall here review what is presently known about the transition in a boundary layer over a smooth

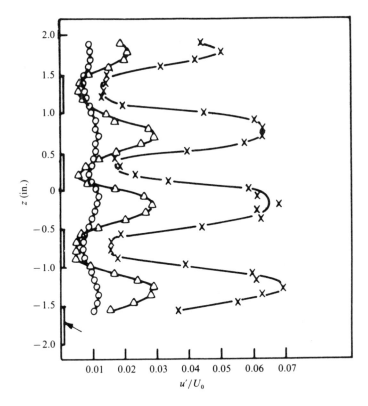

Figure 7.10. Spanwise distributions of intensity of u fluctuation at different distances downstream from vibrating ribbon: 145 c/s wave, $y = 0.042$ in., $U_0/\nu = 3.1 \times 10$ ft^{-1}. $\circ$, $x = 3$ in.; $\triangle$, $x = 6$ in.; $\times$, $x = 7.5$ in. (from Klebanoff, Tidstrom, and Sargent 1962).

flat plate in a flow with a low level of preexisting disturbances. Transition in the presence of surface roughness or of large-amplitude flow disturbances proceeds in a completely different fashion (see Herbert and Morkovin 1980).

Beyond some distance from the leading edge (corresponding to a Reynolds number of about 50,000 based on distance from the leading edge of the flat plate) the Tollmien–Schlichting waves become unstable and begin to grow. How much a given initial disturbance will have grown at a particular downstream distance may be found from the theory described in the preceding section. Semiempirical theories have been proposed (Smith and Gamberoni 1956; van Ingen 1956) that attempt to correlate transition with total linear amplification. They state that transition

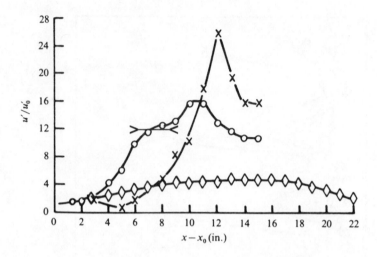

Figure 7.11. Wave growth at peak and valley: 145 c/s wave, $y = 0.045$ in., $U_0/\nu = 3.1 \times 10$ ft^{-1}. Peak: $\wedge$, $u'/U_0 = 0.0008$; $\circ$, $u'/U_0 = 0.00$. Valley: $\vee$, $u'/U_0 = 0.0007$; $\times$, $u'/U_0 = 0.005$. $\succ\!\!\prec$, breakdown (from Klebanoff, Tidstrom, and Sargent 1962).

takes place when the total amplification exceeds about e^9, but different experiments have indicated that the exponent may actually vary between 6 and 12.

When the Tollmien–Schlichting waves have reached an amplitude near 1% of the free-stream velocity, a secondary instability in the form of oblique waves sets in. Only a limited amount of theoretical and experimental information is as yet available as to the characteristics of this instability. Three different types of such instabilities have been identified. By placing spacers (small pieces of Scotch® tape) at regular intervals underneath their vibrating ribbon, Klebanoff, Tidstrom, and Sargent (1962) found that in their experimental situation a spanwise spacing of about 1.3 times the boundary layer thickness produced the strongest spanwise irregularities in streamwise velocity amplitudes downstream of the oscillating ribbon (see Figure 7.10). They found spanwise regions of strong peaks and valleys in the amplitude, the peaks generally occurring halfway between the spacers. The peaks and valleys only occurred when the disturbance amplitude reached a threshold amplitude of about 0.7% of the free-stream velocity (Figure 7.11), indicating that a nonlinear mechanism is at play. Many efforts have been made to explain this spanwise selectivity through nonlinear interactions involving oblique waves (Benney and Lin 1960; Craik 1971; Herbert 1983), but so far none has been found to model

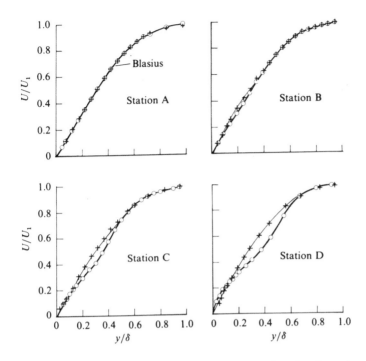

Figure 7.12. Mean velocity distributions across boundary layer at peak and valley: $\circ$, $z = -0.2$ in.; $+$, $z = 0.75$ in. (from Klebanoff, Tidstrom, and Sargent 1962).

the Klebanoff-type secondary instability in a completely satisfactory manner. Craik's (1971) mechanism, which is based on resonance between a two-dimensional and a pair of three-dimensional oblique waves, has been verified in recent experiments by Saric and Thomas (1984), as has Herbert's (1983) mechanism in which the vertical vorticity mode (termed *Squire mode*) plays an essential role. Recent numerical calculations by Orszag and Patera (1983) for a channel flow indicate that such secondary instability may be primarily inviscid.

The three-dimensional secondary instabilities also produce changes in the mean velocity profile. In the experiments by Klebanoff, Tidstrom, and Sargent (1962) the mean velocity profiles in the peak region become inflectional further downstream whereas the ones in the valley regions become fuller (Figure 7.12). The inflectional profiles could be expected to amplify the Tollmien–Schlichting waves more strongly, thus explaining in part the more rapid growth in the peak region. The interaction between the mean and disturbed flow gives rise to spanwise stretching of vorticity, which

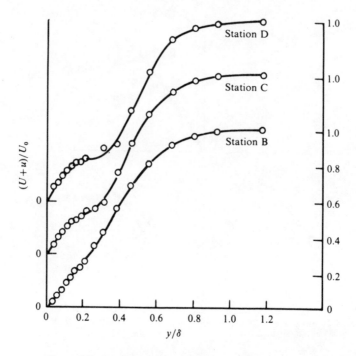

Figure 7.13. Instantaneous velocity distributions across boundary layer: 145 c/s wave, $U_0/\nu = 3.1 \times 10$ ft^{-1} (from Klebanoff, Tidstrom, and Sargent 1962).

produces localized regions of high shear. In these regions the instantaneous velocity distribution shows a marked inflectional character (Figure 7.13) in a thin internal shear layer during part of the cycle. Such regions of very high instantaneous shear were also seen in the experiments by Kovasznay, Komoda, and Vasudeva (1962). A shear layer inside the flow could be expected to be highly unstable on the basis of hydrodynamic stability theory. Inviscid stability theory predicts that the growth rate of an unstable disturbance should be proportional to the local shear and thus be inversely proportional to shear layer thickness. One would thus expect a new ("tertiary") instability to set in once there is substantial inflection in the instantaneous velocity profiles. In the Klebanoff, Tidstrom, and Sargent (1962) experiments this is the case from their station C on (see Figure 7.13). However, strong small-scale instability was observed to occur first some distance downstream at around station D (Figure 7.14), with high-frequency "spikes" accounting for the major increase in disturbance energy. This process, termed *breakdown,* was found to have a distinct frequency (Figures 7.15 and 7.16) of about eight times the fundamental (determined by counting the number of low-velocity spikes per unit time).

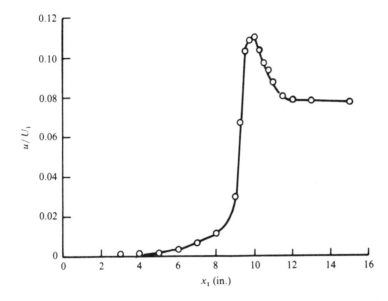

Figure 7.14. Variation of disturbance amplitude with distance from oscillating ribbon (from Klebanoff, Tidstrom, and Sargent 1962).

A number of different theoretical models have been proposed for explaining the breakdown process. Greenspan and Benney (1963) considered the temporal evolution of instability waves on a shear layer with periodically varying thickness and found that there was a substantial intensification of the instability growth during the part of the cycle during which the shear layer was decreasing in thickness. Hocking, Stewartson, and Stuart (1972) analyzed a finite-amplitude wave group propagating on a uniform parallel background shear and were able to show that such a group, modeled on the basis of the nonlinear Schrödinger equation (6.32), could reach infinite amplitudes in a finite time through a self-focusing mechanism. The effects of spatial nonhomogeneity were considered by Landahl (1972). He employed kinematic wave theory to analyze the development of a small-scale, high-frequency instability wave riding on a slowly temporally and spatially varying background flow produced by the primary Tollmien-Schlichting wave after its nonlinear development. (For kinematic wave theory, see Chapter 6.) Since the primary wave grows fairly slowly, its velocity field may be taken as approximately steady in a frame of reference moving with the phase velocity c_0 of the primary wave. The short-wave group will hence travel relative to this frame of reference with a velocity of

$$c_r = c_g - c_0. \tag{7.60}$$

Breakdown of 145 c/s wave ($y = 0.12$ in., $z = -0.2$ in.)

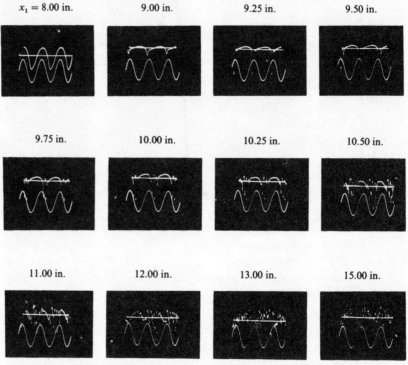

$x_1 = 8.00$ in. 9.00 in. 9.25 in. 9.50 in.

9.75 in. 10.00 in. 10.25 in. 10.50 in.

11.00 in. 12.00 in. 13.00 in. 15.00 in.

Figure 7.15. Oscillograms of u fluctuations illustrating breakdown process: time increasing from left to right, decreasing velocity in downward direction; $U_0/\nu = 3.1 \times 10$ ft^{-1} (from Klebanoff, Tidstrom, and Sargent 1962).

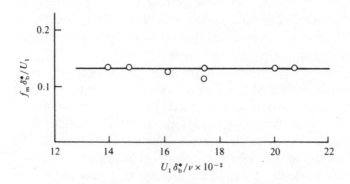

Figure 7.16. Nondimensional plot of "hairpin" eddy frequency (from Klebanoff, Tidstrom, and Sargent 1962).

According to kinematic wave theory (see Chapter 6), wave focusing of an unstable wave will occur in a steady-state system whenever $c_r = 0$, that is, when

$$c_i > 0, \qquad (7.61)$$

together with

$$c_g = c_0 \qquad (7.62)$$

for the wave anywhere along its ray trajectory. At such a point the short wave will also be trapped; that is, the wave group will remain at a position along the primary wave where it can continue to amplify and thus will be absolutely unstable in a frame of reference moving with velocity c_0. The search for the condition for wave breakdown thus amounts to determining the local dispersion relation for the short wave (which generally needs to be done numerically) to see if there is any locally unstable wave for which $c_g = c_0$. Such calculations were carried out (Landahl 1972) using the measured instantaneous velocity profiles in the Klebanoff, Tidstrom, and Sargent (1962) experiment. His results indicate good agreement with the experiments of Klebanoff, Tidstrom, and Sargent (1962), both for the position of breakdown and for the oscillation frequency.

Transition in thermal convection

Transition to chaotic behavior and to fully developed turbulence in thermal convection has been quite extensively studied, both theoretically and experimentally.

As the Rayleigh number is increased beyond its first critical value, successively higher modes of instability arise. The onset of a new instability signifies the occurrence of bifurcation of the nonlinear solution for the convective flow. The linear stability theory predicts the onset of a steady motion in the form of convection rolls. Busse and Clever (1979) have extended the stability analysis to treat secondary instability in the form of oscillations of the convection rolls.

Theoretical modeling of the complete sequence of instabilities and bifurcations leading to transition to fully developed turbulent convection is extremely difficult in view of the complexity of the flow. However, models of the type pioneered by Lorenz (1963) consisting of coupled nonlinear equations for a finite number of discrete modes may provide useful qualitative explanations of the observed features. Curry's (1978) extension to a 14 degrees of freedom model manifests many of the phenomena observed in the detailed experiments of Gollub and Benson (1980) including subharmonic bifurcation and quasi-periodic motion.

8

Shear flow turbulence structure

The structure of shear flow turbulence has been the subject of intense, mostly experimental, studies over many years. The major experimental tools have been hot-wire and laser anemometry supplemented with flow visualization techniques together with statistical sampling methods. The experimental difficulties are great because of the wide range of temporal and spatial turbulence scales present in the flow. Also, interpretation of the data often is difficult, resulting in substantial differences of opinion as to what are the essential features. A popular research subject in recent years has been that of coherent structures, i.e., large-scale, randomly recurring structures with some measure of spatial coherence educed from the turbulent flow field with the application of some suitably chosen conditional sampling technique in order to suppress the background small-scale turbulence. We shall here give a brief review of the published work on turbulence structure of boundary layers as well as of free shear flows, which have been found to differ in important respects. Recently, a new, powerful tool has become available for the study of turbulence and its structure, namely, numerical simulation. Because of its importance and future potential, this tool is treated separately in Chapter 12.

Boundary layer turbulence structure

An important discovery in recent years is that the most active dynamical processes take place in a region very close to the wall, in the region $5 < y^+ < 70$, which at high Reynolds numbers may constitute a very small portion of the boundary layer, often less than 1% of the boundary layer thickness. That there is strong activity in the near-wall region was shown already by Townsend (1956), but the crucial role this region plays in the turbulence dynamics was first conclusively demonstrated by Kline et al. (1967). They used hydrogen bubbles generated by an electric wire to visualize the flow in a low-speed water channel. This technique revealed the

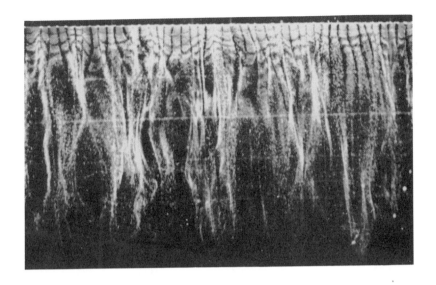

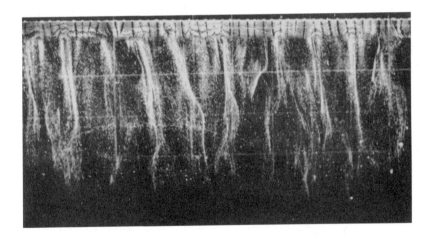

Figure 8.1. Near-wall structure of a turbulent flat-plate boundary layer (top view) visualized with hydrogen bubbles (from Kline et al. 1967).

presence of low-speed streaks (Figure 8.1) of fairly regular spanwise spacing of about $\lambda^+ = 100$ (Figure 8.2). Intermittently, the streaks begin to oscillate and then break up in a fairly violent motion, a "burst" (Figure 8.3). Kline et al. (1967) proposed a conceptual model for bursting, shown in Figure 8.4. In this model the fundamental dynamical mechanisms

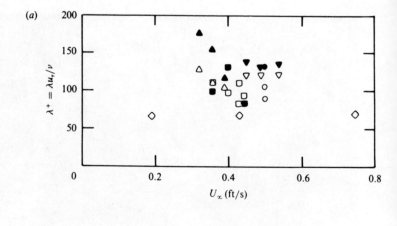

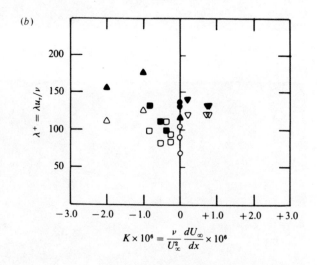

Figure 8.2. Wall streak spacing: (*a*) influence of free-stream velocity; (*b*) influence of pressure gradient. Open symbols, spacing from visual count; closed symbols, obtained from Kim, Kline, and Reynolds (1971).

sorted out are the lift-up of low-speed streak and stretching of spanwise vorticity producing a thin shear layer that is inflectionally unstable. The breakdown of this shear layer, possibly caused by Kelvin–Helmholtz-type instability, would therefore constitute the main source of small-scale turbulence.

Later measurements by Kim, Kline, and Reynolds (1971) showed that much of the turbulence production (about 70% of the total) was associated

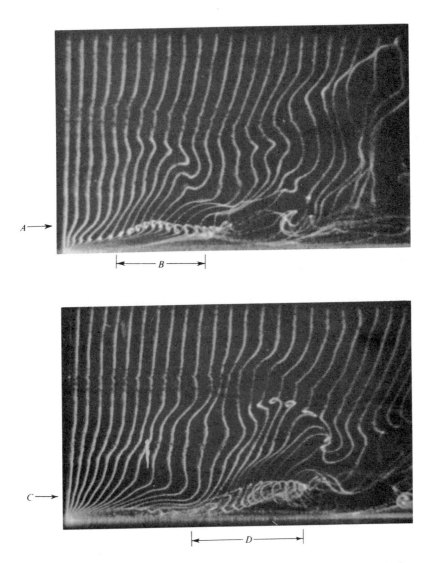

Figure 8.3. Near-wall structure of a turbulent boundary layer visualized by hydrogen bubbles (side view). Sequence showing formation of streamwise vortex during second stage of bursting. Lower photograph shows flow 0.5 s after upper one (from Kline et al. 1967).

with the bursting. Further visual studies by Corino and Brodkey (1969) and others in Brodkey's group showed that there were two kinds of stress-producing motions: ejections involving rapid outflow of low-speed fluid from the wall region and sweeps characterized by inflow of high-speed

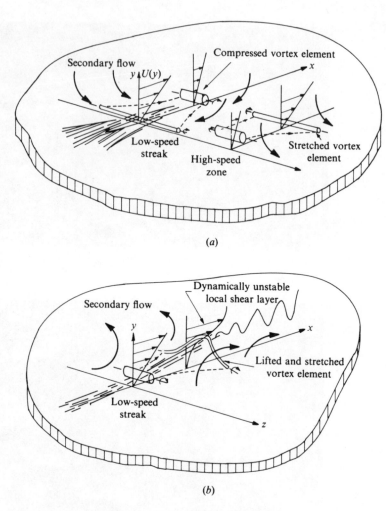

Figure 8.4. Conceptual burst model. (*a*) Mechanics of streak formation, (*b*) mechanics of streak breakup (from Kline et al. 1967).

fluid toward the wall (Figure 8.5). Both of these processes contribute about 70% of the total Reynolds stress. The balance is made up of wall-ward and outward interactions giving negative stress contributions. The ejections could sometimes involve moderately small scale structures, on the order of 20 viscous lengths in width. The stress-producing motions are highly intermittent, occurring maybe only about 25% of the total time.

The appearance of shear layers has been confirmed by many investigations. The shear layer is often referred to as a high-speed front. Such a

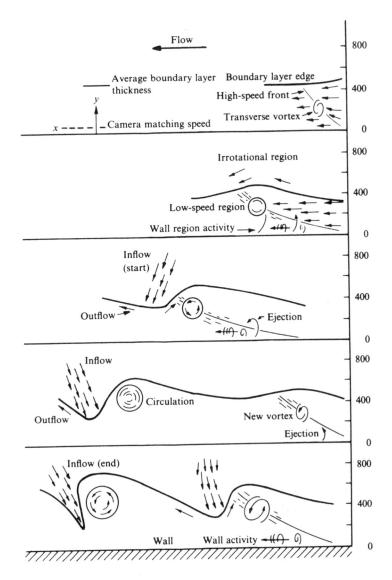

Figure 8.5. Progression of flow illustrating (conceptual) formation of spanwise vortices (from Praturi and Brodkey 1978).

front (or shear layer) moves with the flow and becomes more and more tilted by the mean shear as it travels downstream. Blackwelder and Kaplan (1976) employed a conditional sampling technique called the Variable Interval Time-Averaging (VITA) technique, which brings out the fronts most clearly. This technique employs a short-time variance defined by

$$\text{var}_u(x_i, t, T) = \hat{u}^2(x_i, t, T) - [\hat{u}(x_i, t, T)]^2, \tag{8.1}$$

where circumflex ($\hat{}$) denotes averaging over the selected time T, i.e.,

$$\hat{u}(x_i, t, T) = \frac{1}{T} \int_{t-T/2}^{t+T/2} u(x_i, t_1) \, dt_1. \tag{8.2}$$

Sampling is carried out whenever the variance exceeds a threshold level defined by

$$\text{var}_u > k u_{\text{rms}}^2. \tag{8.3}$$

Here k is a selected threshold level (varied between 0.5 and 2) and u_{rms} is the root-mean-square of the streamwise component. Blackwelder and Kaplan (1976) selected the averaging time to be 10 in viscous units (ν/u_*^2). Their detection probe was located at $y^+ = 15$, near where the turbulence production is maximum (Kline et al. 1967). With a hot-wire rake they measured simultaneously the streamwise velocities at 10 different distances from the wall. By varying the threshold parameter k, they found that the reduced conditionally sampled streamwise velocity

$$\langle u \rangle^* = \frac{\langle u \rangle}{(k u_{\text{rms}}^2)^{1/2}} \tag{8.4}$$

was approximately independent of k. For most of their presented data they chose $k = 1.2$.

As can be seen by using for the velocity a local Taylor series expansion, in time the variance tends to become large for large values of the rate of change of velocity. Hence, the VITA technique tends to single out motions that have high acceleration or deceleration. In their investigation of turbulent channel flow employing the VITA technique, Johansson and Alfredsson (1982) found that the collapse of data with the use of (8.4) became even better if accelerating and decelerating events were treated separately.

The instantaneous velocity distributions obtained in the experiments by Blackwelder and Kaplan (1976) are shown in Figure 8.6. It is seen that there is a substantial y coherence of the data but little z coherence. Conditionally sampled results following (8.1)–(8.4) are shown in Figure 8.7. The time is measured relative to the time at which (8.3) is satisfied. At early times a velocity defect develops near the wall and a velocity excess farther out. The velocity defect near the wall and the excess farther out increase until the detection time (time zero in the figure). Very shortly

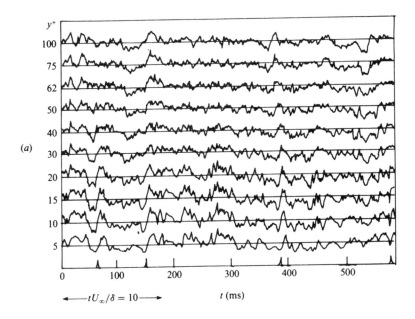

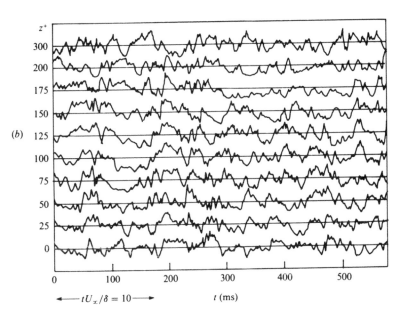

Figure 8.6. Instantaneous streamwise velocities: (a) As function of normal direction (tick marks show satisfaction of detector function at $y^+ = 15$); (b) as function of spanwise coordinate for $y^+ = 15$, $R_\theta = 2550$ (from Blackwelder and Kaplan 1976).

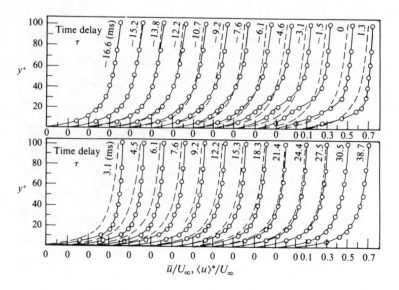

Figure 8.7. Conditionally averaged and mean velocity profiles with positive and negative time delay τ relative to point of detection (from Blackwelder and Kaplan 1976).

after the instant of detection the velocity excess region extends across the whole wall layer and the velocity thereafter relaxes slowly back to the mean value.

The behavior of the conditionally sampled data for the accelerating events may be explained by the intermittent appearance of shear layers, tilted relative to the y axis and swept past the measuring station. The passage of the high-speed front appears as a sweep. The shear layer (front of high-speed region) is highly swept near the wall, as shown in the experiments by Kreplin and Eckelmann (1979). They find that nearest the wall the angle formed by the front with the wall is only about 5° (see Figure 8.8). At larger distances outside the wall region the angle becomes much steeper, about 45° (Head and Bandyopanah 1981).

Several investigators have suggested that the low-speed streaks are associated with longitudinal vortices. An investigation by Blackwelder and Eckelmann (1979) of turbulent channel flow carried out in an oil channel addressed this question. In this flow facility detailed studies of the viscous wall layer are quite feasible because of the large thickness of the wall layer, $y^+ = 17$ corresponding approximately to 1 cm. They found that the low-speed region is formed in between two counterrotating vortices having a streamwise length estimated to be at least $x^+ = 1000$. The longitudinal

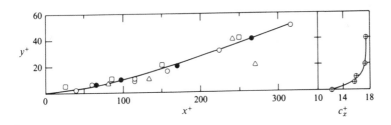

Figure 8.8. Left: averaged front obtained from covariance of streamwise velocity. Right: convection velocity of the front (from Kreplin and Eckelmann 1979).

vorticity was found to be very weak, an order of magnitude smaller than the mean spanwise vorticity.

An associated problem that has received a great deal of attention is the question of how often bursting occurs on the average. The early experiments by Kim et al. (1971) found that the rate of bursting (detected by visual techniques, hence somewhat subjective) scaled with wall variables u_* and ν. Later experiments by Rao, Narasimha, and Badri Narayanan (1971) indicated that outer scaling (based on δ and U_∞) gave best collapse of the data. For channel flow, Johansson and Alfredsson (1982) and Alfredsson and Johansson (1984) showed that if the bursting is detected by use of the VITA technique, the results depend strongly on the threshold, k, and on the averaging time, T, used. In fact, the VITA averaging acts as a low-pass filter with the averaging time closely related to the characteristic time scale of the event (acceleration or deceleration). By plotting the number of events as a function of the VITA-averaging time and taking the maximum, they were able to show that the number of VITA events per unit time best scaled with an intermediate time scale, which is the mean of the outer and wall-related scales.

Theoretical model for wall layer bursting

It is possible to arrive at a qualitative understanding of the near-wall turbulence dynamics with the aid of a simple model which emphasizes the role of the interaction of a turbulent eddy with the strong mean shear in that flow region. The evolution of an isolated three-dimensional eddy of initially weak strength in a parallel shear flow $\bar{U}$ may be thought to occur on three different time scales, namely

(1) shear interaction time $t_s = 1/\bar{U}_w$, ($\bar{U}_w$ = wall shear stress)
(2) viscous interaction time t_v (to be defined later)

(3) nonlinear interaction time scale $t_n = L/u_0$, L being a typical stream-wise length scale of the initial disturbance and u_0 the scale of the initial streamwise disturbance velocity. Of these, t_s is clearly the smallest one in view of the strong shear in the near-wall region of a turbulent boundary layer. Therefore, with the use of matched asymptotic expansions, the asymptotic limit $t/t_s \to \infty$ of the shear interaction solution may be used to determine the inner (in t) limit for both the viscous and nonlinear stages, i.e., to provide the initial conditions for the calculation of these later development stages of eddy evolution. Subdividing the flow field in a mean (taken to be parallel, $\bar{U}(y)$) and a fluctuating one in the usual manner, one has for the inviscid linear stage the equations

$$\frac{\bar{D}u}{\bar{D}t} + \bar{U}'v = -\left(\frac{1}{\rho}\right)\frac{\partial p}{\partial x}, \tag{8.5}$$

$$\frac{\bar{D}v}{\bar{D}t} = -\left(\frac{1}{\rho}\right)\frac{\partial p}{\partial y}, \tag{8.6}$$

$$\frac{\bar{D}w}{\bar{D}t} = -\left(\frac{1}{\rho}\right)\frac{\partial p}{\partial z}, \tag{8.7}$$

$$\frac{\partial u}{\partial x} + \frac{\partial v}{\partial y} + \frac{\partial w}{\partial z} = 0, \tag{8.8}$$

where $\bar{D}/\bar{D}t = \partial/\partial t + \bar{U}\partial/\partial x$, $\bar{U}' = d\bar{U}/dy$, with initial conditions $u = u_0$, $v = v_0$, $w = w_0$ at $t = 0$. The disturbance created by the eddy is assumed to be localized so that the disturbance velocities vanish at large distances. Taking the horizontal divergence of (8.5), (8.7) and using (8.8) one obtains

$$\frac{\nabla_h^2 p}{\rho} = -\frac{\bar{D}v_y}{\bar{D}t} + \bar{U}'v_x, \qquad \nabla_h^2 = \frac{\partial^2}{\partial x^2} + \frac{\partial^2}{\partial z^2}, \tag{8.9}$$

subscripts denoting partial derivatives. An integration of (8.5) over time following a fluid element yields

$$u = u_0(\xi, y, z) - \bar{U}'\mathbf{l} - \frac{\partial \mathbf{P}}{\partial x}, \tag{8.10}$$

where $\xi = x - \bar{U}t$, $\mathbf{l}$ is the liftup of the fluid element, determined by

$$\mathbf{l} = \int v\,\bar{D}t \equiv \int_0^t v(x - \bar{U}(t - t_1), y, z, t_1)\,dt_1, \tag{8.11}$$

and where

$$\mathbf{P} = \int p\,\bar{D}t. \tag{8.12}$$

Integration of (8.9) over time gives for $\mathbf{P}$

$$\frac{\nabla_h^2 \mathbf{P}}{\rho} = v_y + u_{0\xi}(x, y, z) + w_{0z}(\xi, y, z) - \bar{U}'\mathbf{l}_x, \tag{8.13}$$

where it is assumed that $\mathbf{l} = 0$ for $t = 0$ and again continuity has been used. Elimination of the pressure from (8.5) to (8.7) gives the Rayleigh equation for the instability of an inviscid parallel flow

$$\frac{\bar{D}\phi}{\bar{D}t} = \bar{U}''v_x, \tag{8.14}$$

where $\phi = \nabla^2 v$. By integrating this over time one finds

$$\phi = \phi_0(\xi, y, t) + \bar{U}''\mathbf{l}_x, \tag{8.15}$$

which may be handled by standard methods for the Poisson equation to yield the solution for v. For the near-wall region, in which y-derivatives are much greater than x- and z-derivatives (the eddy has a boundary layer character) the following approximation holds (Landahl 1990a):

$$v = \int_0^\infty C(y, y_1)\phi(x, y_1, z, t)\,dy_1$$

$$= \int_0^y (y - y_1)\phi(x, y_1, z, t)\,dy_1 - y \int_0^\infty \phi(x, y_1, z, t)\,dy_1, \tag{8.16}$$

where $C(y, y_1) = (1/2)[|y - y_1| - y - y_1]$. Substitution of (8.16) into (8.11) gives for the liftup $\mathbf{l}$

$$\mathbf{l} = \int_0^\infty C(y, y_1)\frac{F_0(\xi_1, y_1, z) - F_0(\xi, y_1, z)}{\bar{U}(y) - \bar{U}(y_1)}\,dy_1, \tag{8.17}$$

where

$$F_0(x, y, z, t) = \int_{-\infty}^x \phi_0(x_1, y, z)\,dx_1 + \bar{U}''\mathbf{l}, \qquad \xi_1 = x - \bar{U}(y_1)t. \tag{8.18}$$

Provided $\int_{-\infty}^\infty v_0(x_1, y, z)\,dx_1 \neq 0$, $\mathbf{l}$ will tend to a finite value as $t/t_s \to \infty$ and become a function of ξ alone, not of x and t individually. Furthermore, the disturbed region will grow in size linearly with t. From (8.15) and (8.16) it follows, upon substitution of ξ_1 as an integration variable instead of y_1 that v will decay for large times as $1/t$. Hence from (8.13) one sees that $\mathbf{P}$, as well, will become a function of ξ only, from which it follows from (8.10) that u, in the limit as $t/t_s \to \infty$, will tend to a nonvanishing value as a function of the Lagrangian variable ξ (and also of y and z, but not of x and t, individually), as will also the spanwise component w.

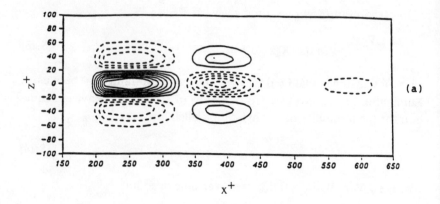

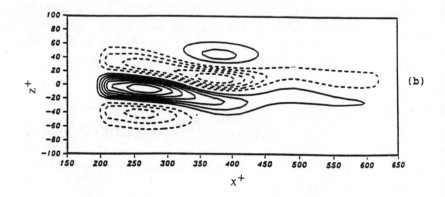

Figure 8.9. Contours of constant streamwise perturbation velocity in the plane $y^+ = 15$ at the nondimensional time $t^+ = 40$ due to an initial disturbance consisting of two counterrotating streamwise rolls: (a) symmetrical structure, $\theta = 0°$; (b) asymmetrical structure, $\theta = 5°$ (from Landahl 1990a).

The disturbed region will grow linearly with time in the streamwise direction. This illustrates the behavior of algebraic instability (Landahl 1980) in which the total streamwise perturbation momentum will grow linearly with time. This instability may also explain the formation of the streamwise streaks of low or high velocity observed in the near-wall region. The example taken from Landahl (1990a) presented in Figure 8.9 shows the evolution of an initial disturbance consisting of two counterrotating streamwise rolls, which have been given a small spanwise asymmetry measured by the angle $\theta = 5°$. The spanwise asymmetry is needed in order to give the initial disturbance a net vertical momentum along lines $y, z = \text{const}$, which is necessary for the algebraic instability to appear. The contours of constant streamwise velocity for large times after the initiation of the disturbance clearly shows a characteristic streaky structure in qualitative agreement with what has been observed in the experiments and in the numerical simulations.

The shear interaction solution for large t/t_s gives an intermediate stage of eddy development which, in addition to producing the streaky structures (Landahl 1990a), reproduces many of the other interesting qualitative flow features seen in the experiments (Kline et al. 1967; Kim et al. 1971) and numerical simulations (Moin and Kim 1982; Johansson et al. 1991). For the streamwise perturbation vorticity $\gamma_1 = \partial w/\partial y - \partial v/\partial z$ one finds (since $v \to 0$) that, in this long-time limit,

$$\gamma_1 = -t\bar{U}'w_{\infty x} + w_{\infty y}, \tag{8.19}$$

subscripts denoting partial derivatives. Similarly for the spanwise component $\gamma_3 = \partial v/\partial x - \partial u/\partial y$,

$$\gamma_3 = \frac{\partial v}{\partial x} - \frac{\partial u}{\partial y} = t\bar{U}'u_{\infty x} - u_{\infty y}. \tag{8.20}$$

For large times the first term in both (8.19) and (8.20) will dominate. Thus, both the streamwise and the spanwise components will intensify for large times. The mechanism responsible for this is the stretching (in case of γ_3) and tilting (in case of γ_1) of the mean spanwise vorticity $-\bar{U}'$ by the gradients of the spanwise perturbation velocity component, w, which in the long-time limit becomes independent of time in a frame of reference moving with the fluid element. Thus, the vorticity will grow linearly with time. The enhancement of spanwise vorticity through this mechanism, together with the rotation by the mean flow, will lead to the formation of a spanwise thin shear layer which is highly tilted in the streamwise direction, as has indeed been found from conditional sampling of experimental and numerical results (Johansson et al. 1991). Similarly, the intensification of streamwise vorticity, as shown by the shear interaction solution, may lead to the formation of concentrated vortices oriented primarily in the streamwise direction. Of course, viscosity, and nonlinearity as well, will eventually limit the shear intensification due to the interaction of the disturbance with the mean shear, as will be seen in the following analysis.

Provided $t_s/t_v \ll 1$ and $t_s/t_n \ll 1$ the limiting results for $t/t_S \to \infty$ for the horizontal perturbation velocity components may now be used as initial values in determining the effects of viscosity and nonlinearity for the long-time evolution of the eddy for $t/t_v = O(1)$ and $t/t_n = O(1)$. Let the limiting values of u and w for $t/t_s \to \infty$ be denoted by u_∞ and w_∞, respectively. The linear equation for u with viscosity included reads

$$\frac{\bar{D}u}{\bar{D}t} + \bar{U}'v = -\left(\frac{1}{\rho}\right)\frac{\partial p}{\partial \xi} + \nu\nabla^2 u. \tag{8.21}$$

For $t/t_s \to \infty$, v and p tend to zero, hence we have the simplified equation to solve

$$\frac{\bar{D}u}{\bar{D}t} = \nu\nabla^2 u, \tag{8.22}$$

which for the inviscid case $\nu = 0$ yields the result $u = u_\infty(\xi, y, z)$, which is consistent with the Taylor (1915) frozen flow field hypothesis. For small viscosity one should therefore try a solution of the form

$$u = f(\xi, y, z, t). \tag{8.23}$$

For the Laplacian one then has

$$\nabla^2 u = f_{\xi\xi} + f_{yy} + f_{zz} - 2\bar{U}'tf_{y\xi} + (\bar{U}'t)^2 f_{\xi\xi}. \tag{8.24}$$

For large times the last term will dominate, hence the equation for u may be approximated as

$$\frac{\partial u}{\partial t} = \nu(\bar{U}'t)^2 f_{\xi\xi}, \tag{8.25}$$

which may be reduced to the ordinary diffusion equation with the aid of the substitution $\tau = \nu(\bar{U}')^2 t^3/3$ yielding

$$\frac{\partial f}{\partial \tau} = f_{\xi\xi}, \tag{8.26}$$

the solution of which, subject to the initial condition $f(\xi, y, z, 0) = u_\infty(\xi, y, z)$, is given by

$$f = \frac{1}{\sqrt{\tau\pi}} \int_0^\tau u_\infty(\xi_1, y, z) \exp\left[-\frac{(\xi - \xi_1)^2}{4\tau}\right] d\xi_1. \tag{8.27}$$

The substitution T gives for the viscous time scale

$$t_v = \left[\frac{L^2}{\nu(\bar{U}')^2}\right]^{1/3}, \tag{8.28}$$

where L is the streamwise length scale of the eddy.

In the treatment of the nonlinear stage of eddy evolution it is assumed that the nonlinear time scale is substantially shorter than the viscous one so that the nonlinear development can be handled by inviscid dynamics. If the converse were true, namely that the nonlinear time scale were much larger than the viscous one, the viscosity would have made the disturbances decay before nonlinearity had time to set in, so that the nonlinear effects would not be interesting. The nonlinear inviscid case is also the primary one to consider for arriving at an understanding of how a turbulent eddy may be regenerated, since in the absence of nonlinearity the eddy motion would eventually decay. For the case of $t_n/t_v = O(1)$ the effects of viscosity and nonlinearity must be treated together, which requires a much more difficult analysis not considered here.

From the behavior of the shear interaction solution for large t/t_s one sees that the pressure induced by the disturbance, equation (8.9), tends to zero in this limit. The horizontal pressure gradients therefore may be neglected in the analysis of that stage of flow evolution. Then, the nonlinear flow evolution may be handled most conveniently in a Lagrangian frame of reference, following Russell and Landahl (1984). Thus, the velocity components are taken to be functions of the material coordinates ξ, η, ζ and time, where ξ, η, ζ are the values of x, y, z at the time $t = 0$ [now being considered as the "outer" nonlinear time, $t/t_n = O(1)$ for which asymptotic large t/t_s behavior has been established]. For the horizontal Eulerian coordinates one finds, when the horizontal pressure gradients are neglected, that

$$x = \xi + t[\bar{U}(\eta) + u_\infty(\xi, \eta, \zeta)], \tag{8.29}$$

$$z = \zeta + t w_\infty(\xi, \eta, \zeta). \tag{8.30}$$

From continuity it follows that, in the transformation from the Lagrangian to the Eulerian system, the Jacobian of the transformation must satisfy

$$J \equiv \frac{\partial(x, y, z)}{\partial(\xi, \eta, \zeta)} = 1, \tag{8.31}$$

which gives

$$\frac{\partial \eta}{\partial y} = x_\xi z_\zeta - x_\zeta z_\xi \tag{8.32}$$

or, with the use of (8.29) and (8.30),

$$y - \eta \equiv 1 = \int_0^\eta \frac{d\eta}{(1 + t u_{\infty\xi})(1 + t w_{\infty\zeta}) - t^2 u_{\infty\zeta} w_{\infty\xi}} - \eta. \tag{8.33}$$

It follows from (8.32, 8.33) that a singularity in y is possible for finite t for points for which

$$(1 + tu_{\infty\xi})(1 + tw_{\infty\zeta}) - t^2 u_{\infty\zeta} w_{\infty\xi} = 0, \qquad (8.34)$$

which would occur at a critical time $t = t_{ns}$, of order t_n, required to satisfy (8.34) in some point in the disturbance flow field. The kinematical interpretation of the condition (8.34) is that the fluid elements converge horizontally to a point forcing an infinite vertical velocity to occur at that point because of continuity. Of course, when the singularity is approached, pressure gradients and viscous stresses will no longer be negligible. How these effects may modify the behavior near the singular point has been considered in Cowley et al. (1990) (see also the references therein), who discuss the use of this theoretical approach for the study of unsteady boundary layer separation. They found that their effects are generally small. Thus, the action of adverse pressure gradients to produce boundary layer separation, which has been proposed as a mechanism involved in sublayer bursting, does not seem to be an important factor in this process.

Particle paths in a turbulent boundary layer near such a kinematic singularity computed from (8.33) (see Landahl 1988) show remarkable similarity to those observed during sublayer bursting in the visualization experiments by Kim et al. (1971).

It is thus seen that the above mechanisms involved in the long-time evolution of a three-dimensional localized disturbance, especially the interaction with the strong mean shear, may give valid qualitative explanations for many of the observed features of the turbulent eddies in the near-wall region during the bursting cycle such as the creation of low- and high-speed streaks, the concentration of spanwise and streamwise vorticity into potentially unstable local shear layers and streamwise vortices, and the creation of local regions of strong vertical velocity (ejections and sweeps) through the horizontal convergence of fluid elements.

Structure of free-shear layers

Flow visualization experiments in a turbulent free-shear layer carried out by Brown and Roshko (1974) showed that the dominant flow structure is a row of transverse vortices with fairly regular spacing (see Figure 8.10). As the shear layer grows in the downstream direction, neighboring vortices may go through a pairing process, much like in the transition process as demonstrated by Winant and Browand (1974). There is also evidence of some small-scale, longitudinal streak formation, but the dominating

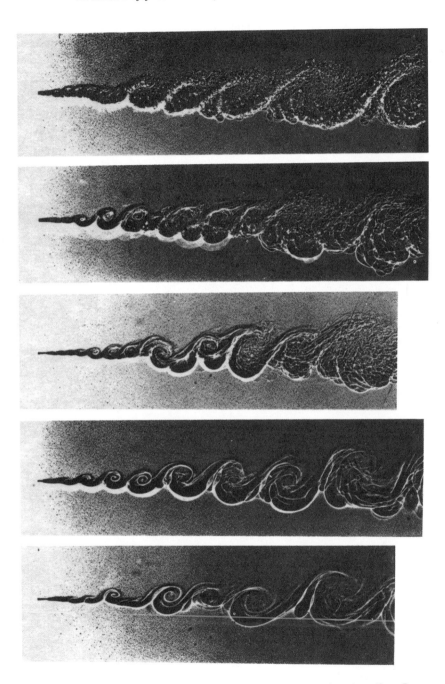

Figure 8.10. Shadowgraphs of mixing layer taken at random times (from Brown and Roshko 1974).

process appears to be an essentially two-dimensional instability of the mean shear (having an inflectional velocity profile), which develops into a row of finite-amplitude vortices going through pairings as the shear layer thickness increases. This view has been challenged by Chadrsuda et al. (1978) on the grounds that the experiments of Brown and Roshko (1974) were carried out in a quiet, ambient flow, and the turbulent flow therefore is likely to contain relics from the transition process, such as vortex pairing. For a shear layer in a turbulent environment they found considerably less two-dimensionality of the turbulence structure and little, if any, evidence of vortex pairing. To study the influence of a large disturbance level, Wygnanski et al. (1979) investigated a turbulent shear layer subjected to forced oscillatory disturbances from a vibrating ribbon. In contrast to Chadrsuda et al. (1978), they found that the two-dimensional vortex structures did dominate even at large forcing amplitudes, thus supporting the initial findings of Brown and Roshko (1974).

9

Turbulence modeling and closure schemes

Because of the difficulties inherent in the nonlinear equations of motion, for theoretical treatment of turbulent flows of practical interest one has so far been forced to resort to approximate and semiempirical prediction methods. For calculation of mean quantities one usually employs what is known as *turbulence modeling,* in which one hypothesizes a functional relationship between the Reynolds stresses and the mean-flow characteristics. This relationship may contain one or more empirical constants to be determined from comparisons with some standard experimental results, for example, the flow over a flat plate. Using the constants thus obtained, one can then predict more complicated flows.

In numerical models of large-scale systems, be they atmospheric models, oceanic models, or models of flow in rivers or around airplanes, one cannot achieve the fine spatial resolution required to describe the small-scale turbulence everywhere. Computer memory size does not suffice at the present stage of computer technology, and technology may never progress that far.

Analogous to the use of bulk variables such as density, temperature, and pressure to describe the macroscopic effects of molecular motions, schemes for describing the effects of turbulence on large-scale flows are in use, although there is room for improvement concerning the representation of the physics of small-scale processes in their effects on larger scales. It would be especially challenging to devise algorithms to describe the large-scale effects that distinguish wave-like, convective, and shear-generated turbulence.

In approximate methods for the statistical theory of turbulence, one tries to relate higher-order statistical quantities to lower-order ones by some *closure scheme* (for example, by relating triple covariances to two-point ones). We shall here review briefly some of the most commonly employed turbulence models and closure schemes.

125

Some turbulence models

The simplest turbulence model is the early one of Boussinesq (1877), who suggested that a turbulent flow could be regarded as having an enhanced viscosity, a turbulent (or eddy) viscosity ν_T. Thus, for the turbulent shear stress

$$\tau_{12} = -\rho\langle uv\rangle = \rho\nu_T\frac{\partial\bar{U}}{\partial y}. \tag{9.1}$$

Generalized for all components the Boussinesq hypothesis gives

$$\tau_{ij} - \tfrac{1}{3}\tau_{kk}\delta_{ij} = \rho\nu_T\left(\frac{\partial\bar{U}_i}{\partial x_j} + \frac{\partial\bar{U}_j}{\partial x_i}\right). \tag{9.2}$$

(The second term on the left is introduced to avoid inconsistencies for $i = j$.) Boussinesq assumed that ν_T is a constant, in which case the equations of mean motion become identical in structure with those for a laminar flow. In this model only one empirical constant, ν_T, needs to be obtained from comparisons and measurements. Mean velocity distributions obtained on the assumption that ν_T is constant have been found to be in good agreement with measurements for simple free-shear flows (jets, wakes), with $\nu_t \approx 10$–100 times the laminar value. However, for wall-bounded flows ν_T must vary with position, since the turbulent fluctuations, and hence ν_T, must be zero at the wall but nonzero in the interior of the flow.

The simplest turbulence model able to account for the variability of the turbulent mixing with the use of only one empirical constant is Prandtl's mixing length model (1925). The basic idea in this model is that a fluid element displaced vertically from its original position y in the boundary layer would retain its original streamwise velocity, the mean velocity $\bar{U}(y)$ at that level. Thus, if the element is displaced vertically a small distance l, its apparent perturbation velocity will be

$$u = \bar{U}(y) - \bar{U}(y+l) \approx -l\frac{\partial\bar{U}}{\partial y}. \tag{9.3}$$

The major physical assumption underlying this hypothesis is that streamwise pressure forces and viscous stresses are unimportant, which may be shown to be justifiable for three-dimensional eddies that are "flat" in the sense that their horizontal dimensions are much greater than their vertical extent (Russell and Landahl 1984). By considering the momentum transfer due to the interchange of two "lumps" of fluid with streamwise perturbation velocities of $-l\,\partial\bar{U}/\partial y$ and $l\,\partial\bar{U}/\partial y$, respectively, Prandtl (1925) reasoned that the v component could be taken as $l|\partial\bar{U}/\partial y|$. Thus, one obtains

$$\tau_{12} = -\rho\langle uv\rangle \approx \rho\langle l^2\rangle \frac{\partial\bar{U}}{\partial y}\left|\frac{\partial\bar{U}}{\partial y}\right| = \rho l_m^2 \frac{\partial\bar{U}}{\partial y}\left|\frac{\partial\bar{U}}{\partial y}\right|, \qquad (9.4)$$

where $l_m = (\langle l^2\rangle)^{1/2}$ is Prandtl's mixing length. For a boundary layer, l_m may be taken to be proportional to the distance y from the wall, that is,

$$l_m = Ay. \qquad (9.5)$$

For the constant-stress region this model gives

$$\mu\frac{\partial\bar{U}}{\partial y} + \rho A^2 y^2\left(\frac{\partial\bar{U}}{\partial y}\right)^2 = \rho u_*^2. \qquad (9.6)$$

Solving this for $\partial\bar{U}/\partial y$, one finds

$$\frac{\partial\bar{U}}{\partial y} = -\frac{\nu}{2A^2 y^2} + \sqrt{(\nu/2A^2 y^2)^2 + u_*^2/A^2 y^2}, \qquad (9.7)$$

or, using wall variables, $U^+ = \bar{U}/u_*$, $y^+ = yu_*/\nu$,

$$\frac{\partial U^+}{\partial y^+} = [\tfrac{1}{2} + \sqrt{\tfrac{1}{4} + A^2(y^+)^2}]^{-1}. \qquad (9.8)$$

For large y^+ this yields

$$\frac{\partial U^+}{\partial y} \approx \frac{1}{Ay^+}, \qquad (9.9)$$

from which it follows that

$$A = \kappa, \qquad (9.10)$$

where $\kappa \approx 0.4$ is von Karman's constant. The velocity distribution in the constant-stress layer may be obtained directly from (9.8) by integration. This gives, with $A = \kappa$,

$$U^+ = \frac{1}{\kappa}\left[\ln(2\kappa y^+ + \sqrt{1+4\kappa^2 y^{+2}}) + \frac{1}{2\kappa y^+}(1 - \sqrt{1+4\kappa^2 y^{+2}})\right]. \qquad (9.11)$$

A large number of more complicated turbulence models have been proposed since Prandtl's initial work. In his own (1945) model he assumed that the eddy viscosity is related to the mean turbulent kinetic energy $q = \tfrac{1}{2}\langle u_i u_i\rangle$ as follows:

$$\nu_T = Aq^{1/2}\Lambda, \qquad (9.12)$$

where A is a constant and Λ is an integral length scale. The turbulent kinetic energy must be found from solving the energy equation (3.28), with additional hypotheses for the transport term and the dissipation term. Jones and Launder (1972) proposed a relation of the form

$$\nu_T = \frac{C_1 q^2}{\epsilon},$$

(9.13)

where C_1 is an empirical constant and ϵ is the dissipation rate. To use this relation, one needs in addition a model equation for the dissipation, as well as the energy equation, suitably modeled. This introduces several new empirical constants and, of course, better possibilities to achieve good fits for many experimental flows. The model proposed by Jones and Launder (1972) is a so-called k-ϵ model (k = kinetic energy), of which several versions exist.

With the development of powerful computers in recent years direct simulation of turbulent flows by numerical integration of the time-dependent Navier–Stokes equations has been considered a realistic possibility. Unfortunately, because of the wide range of temporal and spatial scales that need to be incorporated, no computer is as yet on the horizon that will have the memory capacity and speed to handle high-Reynolds-number turbulent flows of practical interest. An alternative is then to simulate only the largest eddies and model the effects of those that are smaller than the grid employed in the numerical solution. Such an approach was employed by Smagorinsky (1963) in calculation of meteorological flow fields. He used for the subgrid scale modeling

$$\nu_T = (C\Delta)^2 \left[\frac{(\partial \bar{U}_i/\partial x_j + \partial \bar{U}_j/\partial x_i)^2}{2} \right]^{1/2},$$

(9.14)

where C is an empirical constant and Δ is the subgrid length scale. This expression suggests that the eddy viscosity is related to the viscous dissipation in the flow. Smagorinski's subgrid scale modeling has also been used by Moin and Kim (1982) in their numerical study of turbulent channel flow.

Closure schemes for spectral theories

In the early work on spectral theories of turbulence a major preoccupation was to find approximate closure schemes for the statistical equations. These constitute an open set; the equations for two-point covariances involve triple covariances, those for the triple ones quadruple ones, and so

on. Even for the conceptually simplest kind of turbulence, namely, the isotropic and homogeneous one, the theoretical difficulties in treating these
equations have been overwhelming. However, some approximate schemes
for closure have been developed that have been fairly successful in repro-
ducing the isotropic decaying spectra with a minimum of assumptions.

The three-dimensional energy spectrum $E(k, t)$ obeys the following
equation:

$$\frac{\partial E}{\partial t} = T - 2\nu k^2 E. \tag{9.15}$$

Here T is the nonlinear spectral transfer term, which comes from the triple
covariance terms. The specific expression for T in terms of the triple covariances is not needed since this term is going to be modeled. It describes
conservative processes, namely, inertial transfer of kinetic energy from
one wave number to a neighboring one. Therefore,

$$\int_0^\infty T(k)\,dk = 0. \tag{9.16}$$

Several inertial-type models have been proposed for T primarily with the
aim to find results for E valid in the high-wave-number regime.

One of the early proposed models relating T to E was made by Heisenberg (1948). He assumed that the energy transfer to wave numbers higher
than k may be described by an eddy viscosity $\nu_T(k, T)$ acting on eddies of
smaller wave number than k. Accordingly, he set

$$\int_0^\infty T\,dk = -2\nu_T(k, T) \int_0^\infty k^2 E(k, t)\,dk. \tag{9.17}$$

Further, by setting

$$\nu_T = \int_k^\infty f[k, E(k, t)]\,dk, \tag{9.18}$$

he found by dimensional reasoning that

$$f = \alpha \sqrt{\frac{E}{k^3}}, \tag{9.19}$$

where α is a constant. Hence,

$$\nu_T = \alpha \int_k^\infty \sqrt{\frac{E}{k^3}}\,dk.$$

This gives

$$\int_0^k T\,dk = -2\alpha \int_k^\infty \sqrt{\frac{E}{k^3}}\,dk \int_0^k k^2 E\,dk, \tag{9.20}$$

which, when substituted into (9.15), yields an integro-differential equation for E. The solution of this gives the result that, for large k, E drops off as the inverse seventh power of k. By considering spectra of the spatial derivatives of u_i, one can demonstrate, however, that E has to drop off faster than any inverse power of k as k tends toward infinity, so that Heisenberg's hypothesis has a serious flaw.

Kovasznay (1948) instead proposed that for $|\partial E/\partial t| \ll |T|$ (i.e., near equilibrium)

$$T = -\frac{\partial G}{\partial k} = -2\nu k^2 E(k). \tag{9.21}$$

Because of (9.16), the new function G satisfies

$$G = \int_0^k T\,dk = -\int_k^\infty T\,dk \tag{9.22}$$

and is approximately equal to the dissipation rate, ϵ, for small k just beyond the energy containing range $k < k_e$. Kovasznay (1948) assumed that G is a function of E and k only, which on dimensional grounds leads to

$$G = \alpha E^{3/2} k^{5/2}. \tag{9.23}$$

When substituted into (9.21), this yields a separable differential equation having the solution

$$E = \alpha^{-2/3} \epsilon^{2/3} k^{-5/3} \left(1 - \frac{\alpha \nu k^{4/3}}{2\epsilon^{1/3}}\right)^2 \tag{9.24}$$

for $k_e < k < k_1$, where

$$k_1 = \left(\frac{2\epsilon^{1/3}}{\alpha \nu}\right)^{3/4}, \tag{9.25}$$

and $E = 0$ for $k > k_1$.

A different hypothesis also applicable to large k was introduced by Pao (1965), who assumed that

$$G = Ef(\epsilon, k). \tag{9.26}$$

Dimensional reasoning then gives

$$\frac{G}{E} = \alpha^{-1} \epsilon^{1/3} k^{5/3}. \tag{9.27}$$

Introducing this into (9.21), he found

$$\frac{\partial G}{\partial k} = -2\alpha\nu\epsilon^{-1/3}k^{1/3}G, \tag{9.28}$$

which may be solved by separation of variables to give G and hence E from (9.27) as

$$E = \alpha\epsilon^{2/3}k^{-5/3}\exp\left[\frac{-3\alpha(k/k_K)^{4/3}}{2}\right], \tag{9.29}$$

where k_K is the Kolmogorov wave number, $k_K = 2\pi l_K^{-1}$, where η is given by (2.18).

There have been a large number of much more ambitious efforts to affect closure by starting directly from the equations of motion. The most extensive work has been that of Kraichnan (1959), who introduced the so-called direct interaction approximation and developed the idea extensively in a long series of papers thereafter.

10

Aerodynamic noise

The phenomenon of sound generated by a turbulent flow is an example of the excitation of a linear acoustic field by the mostly nonlinear processes of turbulence. This is the central idea in Lighthill's (1952) pioneering paper on aerodynamic sound. Because of the complexities of turbulent flows, most of the research and development has been devoted to better understanding of the source terms for the noise. The additional effects of source motion and advection of sound by the jet flow are largely understood.

In this chapter we first give the simple version of an acoustic field excited by a turbulent flow, then we touch on the effects of moving sources and sound advection, and then, finally, we discuss the dynamics of interactions between vorticity, potential flow disturbances, and density fluctuations added to a mean flow.

A linear acoustic field excited by turbulent fluctuations

Lighthill (1952) started with the equations of motion and manipulated them so as to isolate a wave operator on pressure or density. He regarded the rest of the terms in the particular equation as source terms. We shall follow that procedure.

Write the continuity equation in the form

$$\frac{\partial \rho}{\partial t} + \frac{\partial (\rho u_i)}{\partial x_i} = 0. \tag{10.1}$$

The momentum equation is written as

$$\frac{\partial (\rho u_i)}{\partial t} + \frac{\partial (\rho u_i u_j)}{\partial x_j} + \frac{\partial p}{\partial x_i} - \mu \nabla^2 u_i = 0. \tag{10.2}$$

We combine these two equations by differentiating (10.1) with respect to t and (10.2) with respect to x_i and subtract them, to obtain

132

$$\frac{\partial^2 \rho}{\partial t^2} - \nabla^2 p = \frac{\partial^2 (\rho u_i u_j)}{\partial x_i \, \partial x_j} - \frac{\partial}{\partial x_i} (\mu \nabla^2 u_i). \tag{10.3}$$

We observe that in a linear acoustic field both the density and pressure fluctuations obey a homogeneous wave equation,

$$\frac{\partial^2 \rho}{\partial t^2} - a^2 \nabla^2 \rho = 0. \tag{10.4}$$

Furthermore, for sound of modest intensity, that is, for $\rho'/\bar{\rho} \ll 1$, where ρ' is the density perturbation, pressure and density fluctuations are related to a good approximation by the isentropic relation

$$\frac{dp}{d\rho} = a^2 \approx \frac{p}{\rho'}, \tag{10.5}$$

or $p = a^2 \rho'$. We therefore introduce the equation of state in the form

$$P = P(\rho, S), \tag{10.6}$$

where S is the entropy. This we differentiate with respect to x_i to obtain, with $(\partial P/\partial \rho)_S = a^2$,

$$\frac{\partial P}{\partial x_i} = \left(\frac{\partial P}{\partial \rho}\right)_S \frac{\partial \rho}{\partial x_i} + \left(\frac{\partial P}{\partial S}\right)_\rho \frac{\partial S}{\partial x_i} = a^2 \frac{\partial \rho}{\partial x_i} + \left(\frac{\partial P}{\partial S}\right)_\rho \frac{\partial S}{\partial x_i}. \tag{10.7}$$

Entropy variations are generally small (except in the case of strong shock waves) so that the last term may be neglected. Substitution of this into (10.3) then yields

$$\frac{\partial^2 \rho}{\partial t^2} - a^2 \frac{\partial^2 \rho}{\partial x_i \, \partial x_i} = -\frac{\partial^2 T_{ij}}{\partial x_i \, \partial x_j} + \frac{\partial a^2}{\partial x_i} \frac{\partial \rho}{\partial x_i} - \frac{\partial}{\partial x_i} (\mu \nabla^2 u_i), \tag{10.8}$$

where $T_{ij} = -\rho u_i u_j$ is the *instantaneous Reynolds stress*.

This has given us a wave operator on the density equated to the double divergence of a Reynolds stress tensor T_{ij} plus one term related to temperature fluctuations (since $a^2 = \gamma RT$) and a term containing the viscous stress. These last two terms may be ignored in most cases.

When the noise emitter is a turbulent jet, the source region is of limited size, and outside the jet the right-hand side of (10.8) is zero. So outside the jet (10.8) describes sound propagation in a stationary medium. For a low turbulent Mach number,

$$M_T = \left(\frac{\langle u_i u_i \rangle}{a^2}\right)^{1/2} \ll 1, \tag{10.9}$$

the turbulent fluctuations have very little effect on sound propagation, so that sound can propagate through the turbulent region as if the medium were undisturbed. The process of aerodynamic generation of sound can thus be approximated by

$$a^2 \nabla^2 \rho - \frac{\partial^2 \rho}{\partial t^2} = \frac{\partial^2 T_{ij}}{\partial x_i \, \partial x_j}, \tag{10.10}$$

where $T_{ij} = 0$ outside a *source volume V*.

The far-field sound

Green's function for the wave equation for a source satisfies the equation

$$\left(\frac{\partial^2}{\partial t^2} - a^2 \nabla^2 \right) G(\mathbf{x}, t; \mathbf{y}, t') = \delta(\mathbf{x} - \mathbf{y}) \delta(t - t'), \tag{10.11}$$

where G is found to be (see, e.g., Morse and Feshbach 1953)

$$G(\mathbf{x}, t; 0, 0) = \frac{(4\pi a^2)^{-1} \delta(t - |\mathbf{x}|/a)}{|\mathbf{x}|} = \frac{(4\pi a^2)^{-1} \delta(t_r)}{|\mathbf{x}|}, \tag{10.12}$$

where t_r is the *retarded time,*

$$t_r(\mathbf{x}, \mathbf{y}) = t - \frac{|\mathbf{x} - \mathbf{y}|}{a}.$$

The solutions for a dipole or quadrupole source distribution can be found by differentiation of the source solution. For the problem at hand (10.10) shows that the emitters behave like quadrupoles, and the corresponding solution for the far field is (see Lighthill 1952)

$$\rho(\mathbf{x}, t) = (4\pi a^4)^{-1} \int \int \int_{V_y} \frac{(x_i - y_i)(x_j - y_j)(\partial^2 T_{ij}/\partial t^2)}{|\mathbf{x} - \mathbf{y}|^3} \, dV_y. \tag{10.13}$$

One can further argue that the directional cosines

$$\cos \theta_i = \frac{x_i - y_i}{|\mathbf{x} - \mathbf{y}|}$$

vary very little over the source volume V_y when $\mathbf{x}$ is far away. Furthermore, in the far field, for $\mathbf{x} \gg \mathbf{y}$,

$$\frac{1}{|\mathbf{x} - \mathbf{y}|} \approx \frac{1}{|\mathbf{x}|} = \frac{1}{r}, \tag{10.14}$$

so that in the far field (10.13) may be approximated by

$$\rho(\mathbf{x}, t) \approx \frac{\cos \theta_i \cos \theta_j}{4\pi a^4 r} \int \int \int_{V_y} \frac{\partial^2 T_{ij}(\mathbf{y}, t_r)}{\partial t^2} \, dV_y. \tag{10.15}$$

Note that in the source term we still need to use the retarded time t_r, since otherwise we eliminate the possible cancellation of signals that arrive simultaneously at **x** but that were emitted from different locations **y** within the source volume.

Similarity analysis for the emitted sound

For the particular case we consider, namely the low-Mach-number case, where one can ignore sound refraction by the mean and fluctuating velocity field, and where sound and "turbulence" can be considered separate phenomena, determination of the far-field noise level requires an estimate of the double divergence of the Reynolds stress tensor

$$\frac{\partial^2 T_{ij}}{\partial x_i \, \partial x_j} = -\frac{\partial^2 (\rho u_i u_j)}{\partial x_i \, \partial x_j}. \tag{10.16}$$

We want to look at how the far-field noise level for a jet varies with jet speed and diameter. From the expression for the far field we obtain the time autocovariance of density fluctuations at a point **x** in the far field as

$$\langle \rho(\mathbf{x}, t) \rho(\mathbf{x}, t+\tau) \rangle = (4\pi)^{-2} a^{-8} r^{-2} \cos \theta_i \cos \theta_j \cos \theta_\alpha \cos \theta_\beta$$

$$\times \iiint_{V_y} \iiint_{V_z} \left\langle \frac{\partial^2 T_{ij}[\mathbf{y}, t_r(\mathbf{x}, \mathbf{y})]}{\partial t^2} \frac{\partial^2 T_{\alpha\beta}[\mathbf{z}, t_r(\mathbf{x}, \mathbf{z})]}{\partial t^2} \right\rangle dV_y \, dV_z. \tag{10.17}$$

One can, of course, exclaim over this expression and wonder how one would ever be able to say much about the right-hand side, being a quadruple velocity-space-time covariance integrated twice over a volume. But one usually has much better luck at guessing at the value of an integral than at the details of the integrand.

For an estimate of how the noise scales with the flow parameters, we will assume that T_{ij} scales with ρU^2, where U is the mean speed in the jet. The time scale of fluctuations will be proportional to D/U, where D is the jet diameter. This gives

$$\frac{\partial^2 T_{ij}}{\partial t^2} \approx \frac{\rho_0 U^4}{D^2}. \tag{10.18}$$

The volume of integration V scales with D^3, so the density fluctuations in the far field scale as follows:

$$\langle \rho^2(\mathbf{x}, t) \rangle \approx \frac{\rho_0 U^8 D^6}{a^8 D^4} = \rho_0 M^8 D^2. \tag{10.19}$$

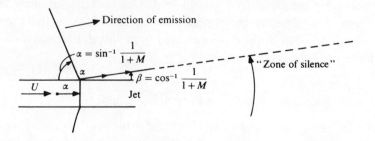

Figure 10.1. Sound waves traveling along a jet.

This is the similarity law found by Lighthill (1952), valid for moderate Mach numbers. A crucial assumption behind this result is that the sound generators are quadrupoles and that they do not change character with changing Mach number, for example, in a manner that generates effective dipole and source emitters because of favorable interference between emitters.

Effects of source motion and sound advection

Equation (10.17) is a Wiener–Khinchine relation, as discussed previously in Chapter 4. But, because of the fact that sound sources in a turbulent flow are moving and cause a Doppler shift, one must somehow include this effect before one can calculate an approximate power spectral density of the far-field sound. This was first done by Ffwcs Williams (1963). He found the expression for the power spectral density of the far-field sound for an axisymmetric jet from a stationary opening in air otherwise at rest to be, approximately,

$$S_{\rho\rho}(\omega, r, \theta) \approx \frac{\rho M^8 (4\pi)^{-2} r^{-2} \phi[\omega/(1 - M\cos\theta)]}{(1 - M\cos\theta)^5}. \tag{10.20}$$

Here $\mathbf{x} = r\cos\theta = x_1/r$, $M = U/a$, U is the mean jet velocity, $F(\theta)$ is an "angular spread" function, and $\phi(\omega)$ is the power spectral density of the emitter. Another effect is that of sound advection by the mean flow in the jet. The essence of this effect is illustrated in Figure 10.1. Sound emitted from a source within the jet will travel in all directions away from the source. The sound that travels in the direction of the jet will travel at a speed of $U + a$ with respect to the air outside. The continuation of a planar sound wave inside a circular jet will be a conical wave with semi-vertex angle α given by

$$\sin \alpha = \frac{a}{U+a} = \frac{1}{1+M}.$$

Advection of sound will therefore tend to spread the emitted noise pattern away from the axis of the jet as has been observed.

Effects of a nonuniform flow

The effect of a nonuniform flow is the really difficult and unsolved problem in jet noise, although much insight has been obtained from observations and experimentation with different configurations and mean velocity distributions. While the theory gives the similarity laws and rough rules for effects of change, our hope for control of jet noise lies in experimentation using qualitative physical insights, possibly supported by semiempirical and numerical models.

While we cannot prescribe how one should go about solving the jet noise problem, there are indications that the early free-shear layer in the jet develops instability fluctuations that are modulated by the large-scale instabilities of the jet as a whole, and, in this interplay between large and small scales, the large-scale strains can distort small-scale turbulence to give large values of the source terms $\partial^2 T_{ij}/\partial x_i\,\partial x_j$. The experiments by Browand and Weidman (1976) indicate this, and other experiments by Michalke and Fuchs (1975) as well as the numerical models of Liu (1974) lend further support to this general idea of intermodulation and distortion. It may prove useful to attempt to explain and/or describe trials of new flow configurations for jet noise alleviation in such terms. Here we shall have to remain content that jet noise control still remains as much an art as a science, but insights are developing that can be helpful to the practitioners of the art.

Two relatively recent reviews of the jet noise problem sum up the directions that have been followed in experimentation and modeling of jet noise during the last decade and more. The review by Ffwcs Williams (1977) discusses a multiplicity of effects that affect noise emission from jet engine installations and jets. The review emphasizes the difficulties of identifying sound sources in experiments. One can, using detector arrays or acoustic focusing mirrors, identify the direction of propagation of sound in the far field. The combined effects of flow refractions and diffractions of sound makes it difficult to identify the sound generator from the apparent far-field direction, partly because sound can trigger shear instability, as shown by Michalke and Fuchs (1975), and because unsteady shear distortion of small-scale turbulence can generate rapidly moving sound generation patterns. Thus, far-field observations alone cannot give much information about the mechanisms of sound generation in a jet.

The second, more recent review by Goldstein (1984) emphasizes the apparently crucial importance of the rapid distortion of small-scale turbulence by large-scale shear fluctuations. The role of this process in shear turbulence has been discussed by Moffatt (1981). Because large-scale shear fluctuations fill most of the cross section of the jet and large-scale disturbances tend to persist for several jet diameters, the sound generated by distortion of small-scale turbulence may also appear to be coherent. A mechanism like this would help explain the early laboratory observations of jet noise by Mollo-Christensen and Narasimha (1960) and by Mollo-Christensen, Kolpin, and Martuccelli (1964). If further experiments should confirm that large-scale shear distortions play a dominant role in certain classes of jet noise, there will be increased observational support for the types of noise modeling suggested by Liu (1974).

Attempts at controlling the large-scale shear fluctuations in the jet by control of initial flow profile and mean initial vorticity structure could then profitably be pursued further.

Instability of a supersonic jet, overreflection

Landau (1944) considered the stability of a discrete supersonic shear layer between uniform parallel flows. He found that it was neutrally stable to small disturbances for Mach-number differences across the layer in excess of $2^{2/3}$. He also mentioned that experiments showed jets in this Mach number range to be unstable. The problem of stability of laterally limited flow, such as a jet, and the relationship between spatial and temporal growth of disturbances can be explained, in part, by considering the process of overreflection of disturbances.

Miles (1957) and Ribner (1957) considered the reflection and transmission of sound waves by a discrete shear layer in a compressible flow. Mollo-Christensen (1959) pointed out the consequences for laterally bounded flows, such as jets, namely the spatial growth of disturbances. Here we will describe how a supersonic shear layer can support sound waves, how this corresponds to overreflection of an incoming disturbance, and the consequent spatial growth of disturbances that also involve sound emission.

Instability of an isolated shear layer

Consider the flow shown in Figure 10.2, where two opposing flows of Mach number $M/2$ create a discrete shear layer at $y = 0$. A stationary

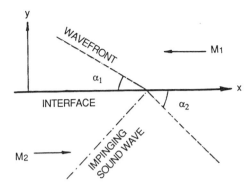

Figure 10.2. Refraction of a sound wave by a shear layer.

bump on the shear layer, given by $y = f(x)$, will emit Mach waves above and below the shear layer. The displacement of the stream lines will be

$$y = f(x + \sqrt{1 - M^2/4})$$

(see Liepmann and Roshko 1957). The corresponding pressure pattern will be symmetric in y, leaving a zero pressure difference across the layer, the pressure at the shear layer being given by (in linear approximation)

$$p(x) = \frac{\rho a^2 M^2/4}{\sqrt{1 - M^2/4}} \frac{df(x)}{dx}. \qquad (10.21)$$

We can now add an impinging wave of zero amplitude, shown as a dotted line in the figure. We take this to show that an impinging wave of zero amplitude will cause finite amplitude reflected and transmitted waves. This implies infinite amplification of the impinging signal, and is an example of singular overreflection. We will not discuss the whole range of instabilities and transmission/reflection properties of compressible shear layers taken up by Landau (1944), Miles (1957), and Ribner (1957); we will simply point out how overreflection can cause spatially growing instability.

Figure 10.3 shows an idealized two-dimensional supersonic jet of finite width h, with a Mach number $M > 2$. A disturbance in the lower layer that moves at half the speed of the jet is neutrally stable and will emit a wave that hits the upper shear layer. The reflected and transmitted waves will be amplified by a factor of infinity according to linear small-perturbation theory. Realistically, it appears that the wave will be amplified, and the reflected wave will impinge on the upper layer, be amplified again, and so on. The strength of the amplification will be limited by the shock relations. Still, we see that a pattern of spatially growing disturbances can

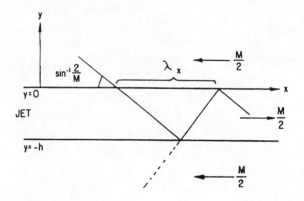

Figure 10.3. Trapping and amplification of an infinitesimal wave by overreflection in a supersonic jet.

be formed, the speed of the pattern being half the speed of the jet, and the direction of the emitted radiation is at an angle of $\cos^{-1}(M/2)$.

Lindzen (1974) and Acheson (1976) have demonstrated the importance of overreflection in geophysical flow stability, and Mollo-Christensen (1977) gave an example involving the horizontal propagation of internal gravity waves. Numerical models of the atmosphere and the ocean may well show that overreflection of large-scale waves may be a significant mechanism in geophysical fluid dynamics if the solutions are analyzed to educe the presence of such processes.

11

Convective transport

The most important manner whereby heat is transported in nature is by turbulent convection. In Chapter 5 we gave as a simple example of thermal convection that between two infinite plates, the lower one of which is heated. The turbulent eddy transport was given by

$$q_3 = \rho c_p \langle w\theta \rangle. \tag{11.1}$$

To determine the average of vertical velocity and temperature fluctuations, the full turbulent flow field is needed, and this is at present not a tractable theoretical problem without resorting to some hypothesis relating the eddy heat flux to the mean temperature. However, a great deal of experimental observations of turbulent convection have been reported in the literature, and we shall review some of the more important ones. When interpreting and studying such data, one is greatly helped by similarity reasoning. Accordingly, we shall make use of some simple similarity laws for turbulent convection. If there is a mean shear flow together with the heating, as one would have in the case of an atmospheric boundary layer, for example, the turbulence will be driven partly by buoyancy, partly by shear, and the flow will influence the transport of heat. Also, the momentum transport will be changed by the convection.

Nondimensional parameters characterizing convective transport

The principal quantities affecting turbulent convection in a fluid layer are the temperature difference across the layer, ΔT, thermal conductivity k, viscosity ν, the reference horizontal velocity U_0, thermal expansion coefficient α, mean density stratification $\bar{\rho}(z)$, and surface roughness height z_0. The quantities desired are the vertical heat flux, $q_3 = q$, the mean velocity distribution, $\bar{U}(z)$, and the wall shear stress, σ_w. From these quantities a number of nondimensional parameters may be formed characterizing the

141

convective motion. We have already considered the Rayleigh number,

$$Ra = \frac{\Delta T d^3 g \alpha}{\kappa_H \nu}, \tag{11.2}$$

which was found to govern the onset of thermal convection, and the Nusselt number,

$$Nu = \frac{qd}{k \Delta T}, \tag{11.3}$$

giving a nondimensional measure of the heat flux. The Nusselt number is a function of the Rayleigh number, as well as of the other nondimensional parameters characterizing the convective flow. The relative magnitude of momentum and heat diffusivity may be measured by the nondimensional Prandtl number,

$$Pr = \frac{\rho c_p \nu}{k}, \tag{11.4}$$

which is solely a material property of the fluid. When horizontal motion is involved, the Reynolds number,

$$Re = \frac{U_0 d}{\nu}, \tag{11.5}$$

enters the picture. For flows in which turbulent convection due to the mean shear flow is dominating, it may be appropriate to express instead the nondimensional heat flux in terms of the Stanton number,

$$St = \frac{q}{\rho_w c_p U_0 \Delta T}, \tag{11.6}$$

where ρ_w is the density at the wall. The Stanton number expresses the ratio of vertical heat flux to the heat carried per unit time by a stream of velocity U_0 and temperature difference ΔT. The Nusselt, Reynolds, and Stanton numbers are related through

$$Nu = Re \cdot Pr_w \cdot St, \tag{11.7}$$

where Pr_w is the Prandtl number at the wall. Additional nondimensional parameters related to the above are the Grashof number,

$$Gr = \frac{Ra}{Pr} \tag{11.8}$$

and the Peclet number,

$$\mathrm{Pe} = \mathrm{Pr} \cdot \mathrm{Re}. \tag{11.9}$$

In flows with density stratification an important parameter is the Richardson number, Ri, which may be taken as an overall value

$$\mathrm{Ri}_0 = -\frac{g\,\Delta\bar{\rho}d}{\rho_0 U_0^2}, \tag{11.10}$$

where $\Delta\bar{\rho}$ is the difference in mean density between upper and lower surface and ρ_0 a reference density, or a local value,

$$\mathrm{Re} = -\frac{g(\partial\bar{\rho}/\partial z)}{\bar{\rho}(\partial\bar{U}/\partial z)^2}. \tag{11.11}$$

The Richardson number gives a measure of the relative effects of buoyancy to those of inertia in the flow and is an important parameter for the stability of a stratified shear flow. When Ri is small (less than $\frac{1}{4}$), the flow may be unstable.

For flows with turbulent heat transport it may be useful also to consider the (local) flux Richardson number,

$$\mathrm{Ri}_f = -\frac{g\bar{\rho}_0 \alpha q}{c_p \tau_{13}(\partial\bar{U}/\partial z)}, \tag{11.12}$$

where

$$\tau_{13} = -\bar{\rho}\langle uw \rangle$$

is the Reynolds shear stress. It follows from the energy equation (3.28) that Ri_f is equal to B/P, the ratio of the production of kinetic energy through buoyancy to that by the workings of Reynolds stresses against the mean shear.

In representing the mean velocity, temperature, and density distributions, we need a suitable reference length for z, of which there are several, besides d. For the viscous wall region near a smooth flat plate, we already introduced the viscous wall length, $l_* = \nu/u_*$, leading to the use of the nondimensional wall distance $z^+ = z/l_*$ in the representation of the mean velocity distribution. For a convection-dominated turbulent flow one could similarly define a heat diffusion wall length given by

$$l_d = \left(\frac{\kappa_H \nu}{g\alpha\,\Delta T}\right)^{1/3}, \tag{11.13}$$

which may be employed to construct a diffusion wall variable

$$z^d = \frac{z}{l_d}. \tag{11.14}$$

In a thermal boundary layer dominated by convection one would expect to find a diffusive layer near the wall for $z^d = O(1)$.

In a boundary layer over a rough surface there is no measurable viscous or diffusive region, and the appropriate nondimensional variable is then

$$\bar{z} = \frac{z}{z_0}. \tag{11.15}$$

For boundary layers for which both heat convection and shear are of importance, one may instead use for nondimensionalization the Monin–Obukhov length L_{M-O} and set

$$\zeta = \frac{z}{L_{M-O}}, \tag{11.16}$$

where L_{M-O} is given by (2.21).

The basic approach in making use of such nondimensional variables is to try to represent the ratio of any mean quantity to a reference value as a universal function of one such variable. Which one to choose depends on the range of physical parameters considered. Fortunately, in most applications of interest (e.g., in meteorology) $l_* \ll z_0$, $l_d \ll z_0$, $z_0 \ll L_{M-O}$, so that a viscous or diffusive layer may be regarded as imbedded in an inertial layer and this in turn imbedded in a thermal layer. In such cases the different regions may be treated separately, the thinner layer serving in a sense as the boundary condition for the next thicker layer.

Thermal convection without mean shear

The most extensively studied problem in thermal convection is the convection in a layer of fluid between two horizontal plates of large dimensions compared to the fluid depth. In Chapter 7 we analyzed the stability of a fluid layer of infinite horizontal extent and found that at a Rayleigh number of 1708 steady convection sets in (for perfectly conducting walls) in the form of an alternating up and down motion with a horizontal wave length λ of about $\lambda/d \approx 2$. Experiments have confirmed the value of the critical Rayleigh number and also that the convection first sets in as rolls of the predicted size. At increasing Rayleigh numbers above the critical, the flow pattern becomes increasingly more complex and eventually fully turbulent, as described in Chapter 7. Measurements of the heat transfer rate as a function of the Rayleigh number carried out by Malkus (1954a, b) showed that the heat flux varies in linear segments with the imposed temperature difference ΔT. He found approximately eight linear segments for Rayleigh numbers ranging from the critical up to about 2×10^6. Above this

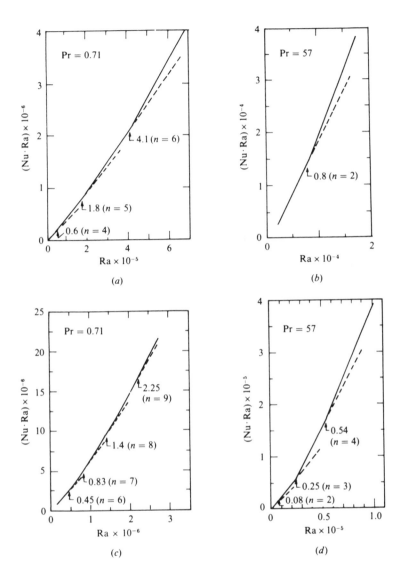

Figure 11.1. Variation of Nu·Ra with Ra (from Willis and Deardorff 1967): (*a*) Pr = 0.71, transitions $n = 4, 5, 6$ are evident; (*b*) Pr = 57, transition $n = 2$ can be seen; (*c*) Pr = 0.71, transitions $n = 6, 7, 8, 9$ are evident; (*d*) Pr = 57, transitions $n = 2, 3, 4$ are evident.

value further discrete changes in the slope were too small to be observed. Malkus interpreted the discrete changes in slope as evidence for transition from one flow regime to the next as successively higher modes become unstable. As an example of such transitions we reproduce in Figure 11.1

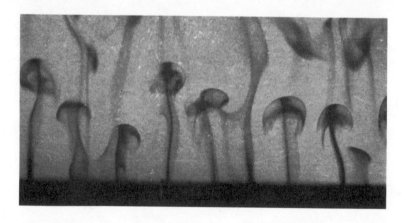

Figure 11.2. Photographs of thermals rising from a heated horizontal surface (from Sparrow et al. 1970).

some results of Nu·Ra as a function of Ra obtained in experiments by Willis and Deardorff (1967). They found that some of the transition points could not be attributed to the onset of linear models, but that nonlinearity must be involved.

At very high Rayleigh numbers (Ra > 10^6) the cellular structure disappears and is replaced by a fully turbulent flow. Then the heat is transported primarily by narrow and isolated thermals that are released from the heated lower surface and penetrate through the well-mixed interior to the upper surface (see Figure 11.2). The mean temperature varies very little with height in the interior. The main temperature variations take place in

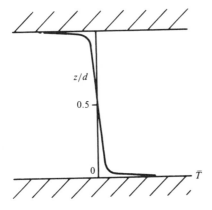

Figure 11.3. Mean temperature distribution across a heated fluid layer at very high Ra (conceptual).

thin thermal boundary layers near the walls, as illustrated schematically in Figure 11.3. In the limit of infinite Rayleigh numbers one would therefore expect the heat flux to be independent of the distance between the plates. Therefore, the Nusselt number

$$\mathrm{Nu} = \frac{q_w d}{k\,\Delta T} \tag{11.17}$$

would be expected to be proportional to the one-third power of the Rayleigh number,

$$\mathrm{Ra} = \frac{g\alpha\,\Delta T\,d^3}{\kappa_H \nu}. \tag{11.18}$$

The relation

$$\mathrm{Nu} \sim \mathrm{Ra}^{1/3} \tag{11.19}$$

was first proposed by Malkus (1954b) and has been found to agree well with experiments at high Ra.

A model for penetrative convection has been proposed by Howard (1966). According to this, thermals are generated by a periodic process in which a heated layer begins to form near the hot wall. As more heat is conducted into this layer, it grows in thickness until its corresponding Rayleigh number exceeds a critical value at which the layer becomes unstable and breaks up, thereby producing a thermal. The mixing associated with the breakup replaces the heated fluid by new cold fluid, and the process starts again. By assuming the breakup phase to be of very short duration compared to the conductive phase, Howard (1966) was able to construct a quantitative theory for the temperature boundary layers and the Nusselt number. The experiments by Sparrow, Husar, and Goldstein (1970) gave cycle times in reasonable agreement with Howard's model and an average Rayleigh number based on the thickness of the conduction layer of about 1800. This is to be compared to the theoretical value of 1100 for the onset of instability in a fluid layer with a linear temperature distribution and a free surface.

12

Numerical simulation of turbulence

An important new tool for studying turbulence has become available in recent years, namely the simulation of turbulent flows by numerical solutions of the Navier–Stokes equations. In some sense one might consider numerical simulations as a new experimental tool, but one in which one has much more control over the experimental conditions than in the laboratory experiment, and one in which more detailed results may be obtained than in the laboratory. Unlike laboratory experiments, numerical simulations do not entail probe size limitations in studies of the smallest scales of motion. Also, in numerical simulation one can obtain information about quantities that are difficult or impossible to measure in the laboratory, such as the instantaneous pressure in the interior of the flow. In this chapter we will give a brief review of the recent developments in this field and the important new understanding of turbulence that has been gained by the use of this tool. For a recent review of this subject we refer to Landahl (1990b).

Numerical techniques

The basic approach in all the numerical schemes that have been developed for the numerical solution of the Navier–Stokes equations for unsteady flows is to use finite difference or spectral decomposition of the (two- or three-dimensional) space for the numerical approximation of the spatial derivatives. From these, the partial time derivatives of u_i are then found, and from them the flow field may be extrapolated to the next time interval. One has to go through a sufficient number of time steps to allow the turbulent field (which is of course basically unsteady) to settle down to a statistically steady state. This may require a total computing time of ten to twenty times the turnover time of the largest dynamically significant eddies.

The computer capacity and speed needed for turbulence simulations may be estimated as follows: The memory requirement is set by the spatial resolution needed to represent the whole range of dynamically significant eddies in the flow. The ratio of the macroscopic (largest) length scale L to the microscopic (smallest) length l may be estimated with the aid of Kolmogorov scaling to be proportional to $R^{3/4}$, where R is the Reynolds number based on L and the root-mean-square fluctuating velocity. From this one finds that the number of points N to resolve the smallest flow structures is proportional to $R^{9/4}$. The number of floating point operations required to update the solution per time step is of the order of $N^{1/3} \log N$. The number of time steps required is found to scale as $R^{1/2}$; therefore, the total number of steps required for one realization is of the order $R^{11/4} \log R$. Such simple estimates give that the computational time needed for simulation of a simple turbulent flow like channel flow on the best available supercomputers (1990 vintage) is of the order one minute for $R = 100$ but several hours for $R = 500$. Reynolds numbers of a few thousands appear to be the practical upper limit today; for ranges of the order of tens of thousands one may have to wait for the next generation of parallel computers. Memory capacity seems to be less of a limitation with supercomputers of the CRAY 2 class. Progress in the research on numerical methods for the solution of Navier–Stokes equations, such as the work by Henshaw et al. (1989), may also lead to grid resolution requirements less severe than obtained from Kolmogorov scaling and thereby allow design of faster computational algorithms.

Some results from numerical simulation

Results from numerical simulations have helped make important progress in the understanding of the transition process as well as of fully developed turbulent flows. Patera and Orzag (1981) were the first to carry out numerical simulations of transition from the two-dimensional Tollmien–Schlichting instability wave stage to the stage of breakdown into small-scale turbulence. They were able to demonstrate that the breakdown is initiated by a secondary three-dimensional instability of the two-dimensional Tollmien–Schlichting wave. A comparison of the growth rate of this instability obtained from their solution with the results of an analysis using modal decomposition is shown in Figure 12.1, which shows good agreement between the two. Other such simulations of transition in wall-bounded flows have been presented by Gilbert (1988) and by Henningson et al.

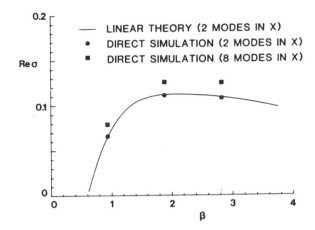

Figure 12.1. The growth rates of three-dimensional disturbances riding on a two-dimensional neutrally stable wave of $\alpha = 1.25$ at $R = 4000$ as function of spanwise wave number β (from Patera and Orzag 1981).

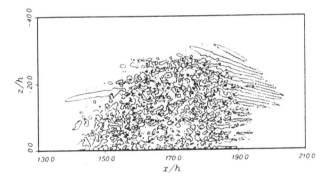

Figure 12.2. Contours of normal velocity at center plane of a turbulent spot in a channel for nondimensional time $t = 258$ (from Henningson et al. 1989).

(1989). In the latter, the evolution of a turbulent spot in a channel flow was computed. The result for the horizontal distribution of the vertical velocity distribution is reproduced in Figure 12.2. Of particular interest is the appearance of the distinct oblique wave trains at the leading edges of the spot; they are also seen in experiments (Lindberg et al., 1984), but are not found in boundary layer spots. The results indicate that the turbulent spot in a channel spreads sideways by the action of oblique instability

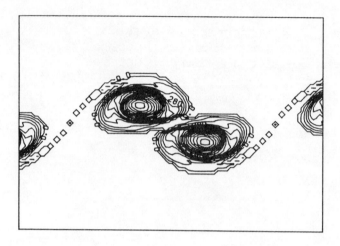

Figure 12.3. (*a*) Spanwise vorticity contours for a free shear layer during vortex pairing. (For times and other parameters for the calculation, see figure 4 in the original paper by Metcalf et al. 1987.)

waves propagating into the flow region outside the spot, which in turn is destabilized by the blocking action of the presence of the traveling and spreading turbulent region (Widnall 1984).

For a free shear layer extensive calculations of the transition process have been carried out by Metcalf et al. (1987). Some of their results are shown in Figure 12.3. The row of vortices created by the roll-up of instability waves are subject to pairing through a subharmonic instability

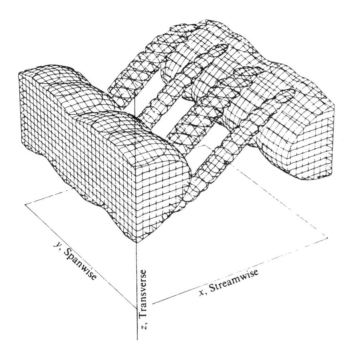

Figure 12.3. (*b*) Three-dimensional plot of 50% level of the sum of the absolute values of all three vorticity components for the shear layer. (For parameters of the run, see figure 10 in the original paper by Metcalf et al. 1987.)

mechanism (Kelly 1967). At a later stage a three-dimensional braided structure of streamwise vorticity appears, possibly as a consequence of secondary instability of the Pierrehumbert–Widnall (1982) type. The braided structure is also seen in fully developed free shear layer turbulence and in turbulent jets (see the review article by Hussain 1980).

The numerical simulations of fully developed turbulent shear flows that have been carried out have given valuable insights into the mechanisms controlling turbulence. Two-dimensional turbulence is of interest in geophysics, since the rotation of the earth tends to inhibit variations in the direction of the rotation axis (the Taylor–Proudman effect). In two-dimensional turbulence there is also a reverse cascade of the enstrophy (the root-mean-square of the vorticity). Figure 12.4 shows results from such a simulation from Henshaw et al. (1989). Particularly interesting is the tendency of the turbulence to become spatially "spotty" for longer times, with localized regions of high intensity separated by large regions of little activity, a property seen in many turbulent flows.

Figure 12.4. Two-dimensional turbulence simulated by Henshaw et al. (1989). Left, $t = 0$. Right, $t = 25.01$.

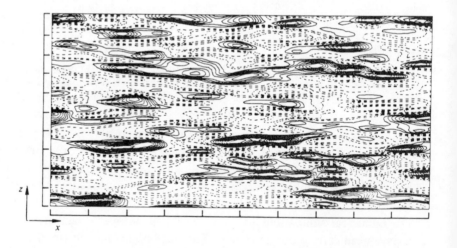

Figure 12.5. Contours of constant streamwise velocity for a turbulent channel flow in the plane $y^+ = 6.14$ (from Moin and Kim 1982).

Extensive numerical simulations of wall-bounded turbulent shear flows have been carried out during recent years by a group at NASA/Ames (Moin and Kim 1982; Spalart 1988). Figure 12.5 shows the contours of constant streamwise velocity in a plane near the wall obtained in the simulations by Moin and Kim (1982). The streaky structure of the velocity

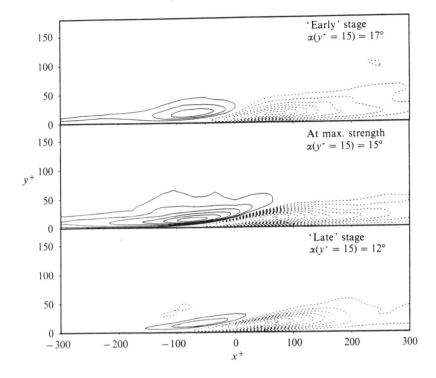

Figure 12.6. (*a*) VISA-educed conditional averages of the streamwise velocity in the *xy*-midplane of the structure at three stages of development (for parameters see original paper of Johansson et al. 1991).

field is quite apparent and in good agreement with the results from the visual experiments of Kline et al. (1967) and Kim et al. (1971).

The data bases generated at NASA/Ames have been generously made available to the whole research community for the use in extensive studies during a series of summer programs (1987–1990) organized by NASA/ Ames and Stanford University (1987, 1988, 1990). As an example of what can be learned about the turbulence structure from such studies we present in Figure 12.6 results obtained by Johansson et al. (1991). They used the VISA conditional sampling techniques on the computed channel flow velocity fields (VISA is the spatial correspondence to VITA, in which streamwise averaging rather than time averaging is used). The conditionally sampled results demonstrate the appearance and intensification of internal shear layers [Figure 12.6(*a*)] and the streamwise elongation and spanwise wiggliness of the high- and low-speed regions that are revealed

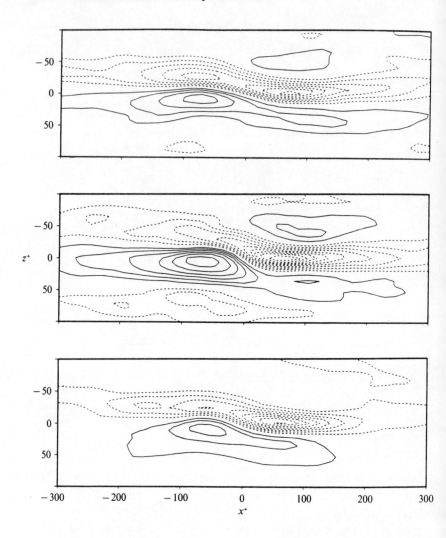

Figure 12.6. (*b*) Conditional averages obtained with a modified VISA-method preserving spanwise asymmetry. Contours of the fluctuating streamwise velocity in the plane $y^+ = 15$ at three stages of development (from Johansson et al. 1991).

when the conditional sampling is carried out such that spanwise asymmetry is retained [Figure 12.6(*b*)]. Landahl (1990a) has shown with the aid of a simple, nearly linear model that the streak formation could be the consequence of algebraic instability (Landahl 1980).

References

Acheson, D. J. 1976. On over-reflexion. *J. Fluid Mech. 77*, 433–72.

Alfredsson, P. H. and Johansson, A. V. 1984. On the detection of turbulence generating events. *J. Fluid Mech. 139*, 325–45.

Bark, F. H. 1975. On the wave structure of the wall region of a turbulent boundary layer. *J. Fluid Mech. 70*, 229–50.

Batchelor, G. K. 1953. *The theory of homogeneous turbulence.* Cambridge University Press, Cambridge.

1967. *An introduction to fluid dynamics.* Cambridge University Press, Cambridge.

Benjamin, T. B. and Feir, J. E. 1967. The disintegration of wavetrains on deep water. Part 1. Theory. *J. Fluid Mech. 27*, 417–30.

Benney, D. J. and Lin, C. C. 1960. On the secondary motion induced by oscillations in a shear flow. *Phys. Fluids 3*, 656–7.

Bernal, L. P. 1981. *The coherent structure of turbulent mixing layers. I. Similarity of the primary vortex structure. II. Secondary streamwise vortex structure.* Ph.D. thesis, California Institute of Technology, Pasadena.

Bers, A. 1975. Linear waves and instabilities. In *Physique des plasmas* (C. DeWitt and J. Peyraud, eds.). Gordon and Breach, New York, pp. 117–215.

Blackwelder, R. F. and Eckelmann, H. 1979. Streamwise vortices associated with the bursting phenomenon. *J. Fluid Mech. 94*, 577–94.

Blackwelder, R. F. and Kaplan, R. E. 1976. On the wall structure of the turbulent boundary layer. *J. Fluid Mech. 76*, 89–112.

Bliven, L. E., Huang, N. E., and Long, S. R. 1986. Experimental study of the effect of wind on Benjamin–Feir instability. *J. Fluid Mech. 162*, 237–60.

Boussinesq, J. 1877. Essai sur la theorie des eaux courantes. *Mem. pres. par div. savants a l'Academie Sci.,* Paris, *23*, 1–680.

1897. *Theorie de l'écoulement tourbillonnant et tumultueux des liquides dans les lits rectilignes a grande section,* 2 vols., Gautier-Villars, Paris.

Browand, F. K. and Weidman, P. D. 1976. Large scales in the developing mixing layer. *J. Fluid Mech. 76*, 127–44.

Brown, G. L. and Roshko, A. 1974. On density effects and large structures in turbulent mixing layers. *J. Fluid Mech. 64*, 775–816.

Busse, F. H. and Clever, R. M. 1979. Instabilities of convection rolls in a fluid of moderate Prandtl number. *J. Fluid Mech. 91*, 319–35.

Chadrsuda, C., Mehta, R. D., Weir, A. D., and Bradshaw, P. 1978. Effect of free-stream turbulence on large structure in turbulent mixing layers. *J. Fluid Mech. 85*, 693–704.

Chandrasekhar, S. 1961. *Hydrodynamic and hydromagnetic stability.* Clarendon Press, Oxford.

Chereskin, T. K. and Mollo-Christensen, E. 1985. Modulational development of nonlinear gravity wave groups. *J. Fluid Mech. 151*, 337–65.

Chin, W. C. 1980. Effect of dissipation and dispersion on slowly varying wavetrains. *AIAA J. 18*, 149–58.

Chu, V. H. and Mei, C. C. 1970. On slowly varying Stokes waves. *J. Fluid Mech. 41*, 873–87.

Clauser, F. H. 1956. The turbulent boundary layer. *Adv. Appl. Mech. 4*, 1–51.

Coles, D. 1956. The law of the wake in the turbulent boundary layer. *J. Fluid Mech. 1*, 191–226.

Corino, E. R. and Brodkey, R. S. 1969. A visual investigation of the wall region in turbulent flow. *J. Fluid Mech. 37*, 1–30.

Cowley, S. J., Van Dommelen, L. L., and Lam, S. T. 1990. On the use of Lagrangian variables in descriptions of unsteady boundary layer separation. ICASE Rep. No. 90-47.

Cramér, H. 1946. *Mathematical methods of statistics.* Princeton University Press, Princeton.

Craik, A. D. D. 1971. Non-linear resonant instability in boundary layers. *J. Fluid Mech. 50*, 393–413.

Curry, J. H. 1978. A generalized Lorenz system. *Comm. Math. Phys. 50*, 193–204.

Dorman, C. E. and Mollo-Christensen, E. L. 1973. Observation of the structure of moving gust patterns over a water surface ("Cat's paws"). *J. Phys. Oceanogr. 73*, 120–32.

Favre, A., Gaviglio, J., and Dumas, R. 1953. Quelques mesures de correlation dans le temps et l'espace en soufflerie. *La Rech. Aero. 32*, 21–8.

Feigenbaum, M. J. 1980. The transition to aperiodic behavior in turbulent systems. *Commun. Math. Phys. 77*, 65–86.

Fermi, E., Pasta, J., and Ulam, S. 1962. Studies of nonlinear problems. In *Collected papers of Enrico Fermi,* Vol. 2. Chicago University Press, Chicago, p. 978.

Ffwcs Williams, J. E. 1963. The noise from turbulence convected at high speed. *Philos. Trans. R. Soc. London, Ser. A 255*, 469–503.

1977. Aeroacoustics. *Ann. Rev. Fluid Mech. 9*, 447–68.

Fjørtoft, R. 1950. Application of integral theorems in deriving criteria of stability of laminar flow and for the baroclinic circular vortex. *Geofys. Publ. 17* (6), 1–52.

Gaster, M. 1962. A note on the relation between temporally increasing and spatially increasing disturbances in hydrodynamic stability. *J. Fluid Mech. 14*, 222–4.

Gilbert, N. 1988. Numerische Simulation der Transition von der laminaren in die turbulenten Kanalströmung. Forshungsbericht DFVLR-FB 88-55, Deutsche Forschungs und Versuchsantalt für Luft- und Raumfahrt, Göttingen, Germany (Doctoral Dissertation at the Universität Karlsruhe).

Goldstein, M. E. 1984. Aeroacoustics of turbulent shear flows. *Ann. Rev. Fluid Mech. 16,* 263–85.

Gollub, J. P. and Benson, S. V. 1980. Many routes to turbulent convection. *J. Fluid Mech. 100,* 449–70.

Greenspan, H. P. and Benney, D. J. 1963. On shear-layer instability, breakdown and transition. *J. Fluid Mech. 15,* 133–53.

Gurvich, A. S., Koprov, B. M., Tsvang, L. R., and Yaglom, A. M. 1967. Data on the small-scale structure of atmospheric turbulence. In *Atmospheric turbulence and radio wave propagation* (A. M. Yaglom and V. I. Tatarskii, eds.). Nauka Press, Moscow, pp. 30–52.

Head, M. R. and Bandyopanah, P. 1981. New aspects of turbulent boundary-layer structure. *J. Fluid Mech. 107,* 297–338.

Heisenberg, W. 1948. On the theory of statistical and isotropic turbulence. *Proc. R. Soc. London, Ser. A 195,* 402–6.

Helleman, R. H. G. 1980. Self-generated chaotic behavior in nonlinear mechanics. In *Fundamental problems in statistical mechanics,* Vol. 5 (E. G. D. Cohen, ed.). North Holland, Amsterdam, pp. 165–233.

Henningson, D. S., Spalart, P., and Kim, J. 1989. Numerical simulation of turbulent spots in Plane Poiseulle and boundary layer flows. *Phys. Fluids A., 30,* 2914–17.

Henshaw, W. D., Kreiss, H. O., and Reina, L. G. 1989. On the smallest scale for the incompressible Navier–Stokes equations. In *Theoretical and computational fluid dynamics.* Springer, Berlin, pp. 65–95.

Herbert, T. 1983. Subharmonic three-dimensional disturbances in unstable plane shear flows. AIAA Rep. No. 83–1759.

Herbert, T. and Morkovin, M. V. 1980. Dialogue on bridging some gaps in stability and transition research. In *Laminar-turbulent transition* (R. Eppler and H. Fasel, eds.). Springer, Berlin, pp. 47–72.

Hinze, J. O. 1975. *Turbulence,* 2nd ed. McGraw-Hill, New York.

Hocking, L. M., Stewartson, K., and Stuart, J. T. 1972. A nonlinear stability burst in plane parallel flow. *J. Fluid Mech. 51,* 705–35.

Howard, L. N. 1966. Convection of high Rayleigh number. In *Proceedings of the eleventh international congress of applied mechanics* (H. Görtler, ed.). Springer, Berlin, pp. 1109–15.

Huang, N. E., Long, S. R., Tung, C.-C., Yuan, Y., and Bliven, L. F. 1981. A unified two-parameter wave spectral parameter model for a general sea state. *J. Fluid Mech. 112,* 203–24.

  1983. A non-Gaussian statistical model for surface elevation of nonlinear random wave fields. *J. Geophys. Res. 88,* 7597–606.

Huang, N. E., Long, S. R., and Bliven, L. F. 1984. The non-Gaussian joint probability density function of slope and elevation for a nonlinear gravity field. *J. Geophys. Res. 89,* 1961–72.

Hui, W. H. and Hamilton, J. 1979. Exact solution of a three-dimensional nonlinear Schrödinger equation applied to gravity waves. *J. Fluid Mech. 93,* 117–33.

Hussain, A. K. M. F. 1980. Coherent structures and studies of perturbed and unperturbed jets. In *The Role of Coherent Structures in Modelling Turbulence and Mixing, Lecture Notes in Physics, 136,* 252–91. Springer, Berlin.

Jimenez, J. and Whitham, G. B. 1976. An averaged Lagrangian method for dissipative wavetrains. *Proc. R. Soc. London, Ser. A. 349*, 277–87.

Johansson, A. V. and Alfredsson, P. H. 1982. On the structure of turbulent channel flow. *J. Fluid Mech. 122*, 295–314.

Johansson, A. V., Alfredsson, P. H., and Kim, J. 1991. Evolution and dynamics of shear-layer structures in near-wall turbulence. *J. Fluid Mech. 224*, 579–99.

Jones, W. P. and Launder, B. E. 1972. The prediction of laminarisation with a 2-equation model of turbulence. *Int. J. Heat Transf. 15*, 301–14.

Kelly, R. E. 1967. On the stability of an inviscid shear layer which is periodic in space and time. *J. Fluid Mech. 27*, 657–89.

Kim, H. T., Kline, S. J., and Reynolds, W. C. 1971. The production of turbulence near a smooth wall in a turbulent boundary layer. *J. Fluid Mech. 50*, 133–60.

Klebanoff, P. 1954. Characteristics of turbulence in a boundary layer with zero pressure gradient. NACA TN 3178.

Klebanoff, P. S. and Diehl, F. W. 1951. Some features of artificially thickened, fully developed, turbulent boundary layers with zero pressure gradient. NACA TN 2475.

Klebanoff, P. S., Tidstrom, K. D., and Sargent, L. H. 1962. The three-dimensional nature of boundary layer instability. *J. Fluid Mech. 12*, 1–34.

Kline, S. J., Reynolds, W. C., Schraub, F. A., and Runstadler, P. W. 1967. The structure of turbulent boundary layers. *J. Fluid Mech. 30*, 741–73.

Kolmogorov, A. N. 1941. The local structure of turbulence in incompressible viscous fluid for very large Reynolds numbers. *Dokl. Akad. Nauk SSSR 30*, 299–303.

Kovasznay, L. S. G. 1948. Spectrum of locally isotropic turbulence. *J. Aeronaut. Sci. 15*, 745–53.

Kovasznay, L. S. G., Komoda, H. S., and Vasudeva, B. R. 1962. Detailed flow field in transition. *Proceedings of the 1962 Heat Transfer and Fluid Mechanics Institute*, Vol. 1. Stanford University Press, Stanford, Calif., pp. 1–26.

Kraichnan, R. H. 1959. The structure of isotropic turbulence at very high Reynolds numbers. *J. Fluid Mech. 5*, 497–543.

Kreplin, H. P. and Eckelmann, H. 1979. Propagation of perturbations in the viscous sublayer and adjacent wall region. *J. Fluid Mech. 95*, 305–22.

Lamb, H. 1932. *Hydronamics,* 6th ed. Dover, New York, p. 417.

Landahl, M. T. 1962. On the stability of a laminar incompressible boundary layer over a flexible surface. *J. Fluid Mech. 13*, 609–32.

  1967. A wave-guide model for turbulent shear flow. *J. Fluid Mech. 29*, 441–57.

  1972. Wave mechanics of breakdown. *J. Fluid Mech. 56*, 775–802.

  1975. Wave breakdown and turbulence. *SIAM J. Applied Math. 28*, 735–56.

  1980. A note on an algebraic instability of inviscid parallel shear flows. *J. Fluid Mech. 98*, 243–51.

  1982. The application of kinematic wave theory to wave trains and packets with small dissipation. *Phys. Fluids 25* (9), 1512–16.

  1988. Linear and non-linear mechanisms in boundary layer turbulence. *International J. for Num. Methods in Fluids 8*, 1183–93.

  1990a. On sublayer streaks. *J. Fluid Mech. 212*, 593–614.

1990b. CFD and turbulence. Daniel and Florence Guggenheim Memorial Lecture. Presented at the 17th International Congress of the Aeronautical Sciences, Stockholm, Sweden.

Landau, L. D. 1944. 'Turbulence. *Dokl. Akad. Nauk SSSR 44* (8), 339–42.

Laufer, J. 1950. Some recent measurements in a two-dimensional turbulent channel. *J. Aeronaut. Sci. 17*, 277–87.

1951. Investigation of turbulent flow in a two-dimensional channel. NACA, Rep. No. 1053.

1954. The structure of turbulence in fully developed pipe flow. NACA, Rep. No. 1174.

Liepmann, H. and Roshko, A. 1957. *Elements of gas dynamics.* Wiley, New York.

Lighthill, M. J. 1952. On sound generated aerodynamically. I. General theory. *Proc. R. Soc. London, Ser. A 211* (1107), 564–87.

Lindberg, P. A., Fahlgren, E. M., Alfredsson, P. H., and Johansson, A. V. 1984. An experimental study of the structure and spreading of turbulent spots. In *Laminar-turbulent transition* (V. V. Kozlov, ed.). Springer, Berlin, pp. 617–24.

Lindzen, R. S. 1974. Stability of a Helmholtz velocity profile in a continuously stratified Boussinesq fluid – applications to clear-air turbulence. *J. Atmos. Sci. 31*, 1507–44.

Liu, J. T. C. 1974. Developing large-scale wavelike eddies and the near jet noise field. *J. Fluid Mech. 62*, 437–64.

Longuet-Higgins, M. S. 1978. The instability of gravity waves of finite amplitude in deep water. II. Subharmonics. *Proc. R. Soc. London, Ser. A 360*, 489–505.

Lorenz, E. N. 1963. Deterministic nonperiodic flow. *J. Atmospher. Sci. 20*, 134–41.

Malkus, W. V. R. 1954a. Discrete transitions in turbulent convection. *Proc. R. Soc. London, Ser. A 225* (1161), 185–95.

1954b. The heat transport and spectrum of thermal turbulence. *Proc. R. Soc. London, Ser. A 225* (1161), 196–212.

Manneville, P. and Pomeau, Y. 1980. Different ways to turbulence in dissipative dynamic systems. *Physica D1*, 219–26.

Melville, W. K. 1983. Wave modulation and breakdown. *J. Fluid Mech. 128*, 489–506.

Metcalf, R. W., Orzag, S. A., Brachet, M. E., Menon, S., and Riley, J. 1987. Secondary instability of a temporally growing mixing layer. *J. Fluid Mech. 184*, 207–44.

Michalke, A. and Fuchs, H. V. 1975. On turbulence and noise of an axi-symmetric shear flow. *J. Fluid Mech. 70*, 179–205.

Miles, J. W. 1957. On the reflection of sound at an interface of relative motion. *J. Acoust. Soc. Am. 29*, 226–8.

Millikan, C. B. 1939. A critical discussion of turbulent flows in channels and circular tubes. *Proceedings of the fifth international congress of applied mechanics* (Cambridge, Mass., 1939), Wiley, New York, pp. 386–92.

Millionschikov, M. 1939. Decay of homogeneous isotropic turbulence in a viscous incompressible fluid. *Dokl. Akad. Nauk SSSR 22*, 236–40.

1941. On the theory of homogeneous isotropic turbulence. *Dokl. Akad. Nauk SSSR 32*, 611–14.

Moffatt, H. K. 1981. Some developments in the theory of turbulence. *J. Fluid Mech. 106*, 27–47.

Moin, P. and Kim, J. 1982. Numerical investigation of turbulent channel flow. *J. Fluid Mech. 118*, 341–77.

Mollo-Christensen, E. 1959. Acoustical instability of jets and wakes for Mach numbers above two. *J. Aero/Space Sci. 26*, 765–6.

1977. Over-reflection of horizontally propagating gravity waves by a vertical shear layer. *Phys. Fluids 21*, 1908–11.

Mollo-Christensen, E. L. and Narasimha, R. 1960. Sound emission from jets at high subsonic velocities. *J. Fluid Mech. 8*, 49–60.

Mollo-Christensen, E. and Ramamonjiarisoa, A. 1978. Modeling the presence of wave groups in a random wave field. *J. Geophs. Res. 83*, 4117–22.

1982. Subharmonic transition and group formation in a wind wave field. *J. Geophys. Res. 87*, 5699–717.

Mollo-Christensen, E. L., Kolpin, M. A., and Martuccelli, J. R. 1964. Experiments on jet noise, far field spectra and directivity patterns. *J. Fluid Mech. 18*, 285–301.

Monin, A. S. 1978. On the nature of turbulence. *Sov. Phys. Usp. 21* (5), 429–42.

Monin, A. S. and Obukhov, A. M. 1953. Dimensionless characteristics of turbulence in the atmospheric surface layer. *Dokl. Akad. Nauk SSSR 93* (2), 223–6.

Monin, A. S. and Yaglom, A. M. 1972. *Statistical fluid mechanics* (2 vol.). MIT Press, Cambridge, Mass.

Moon, H. T., Huerre, P., and Redekopp, L. G. 1983. Transition to chaos in the Ginzburg–Landau equation. *Physica 7D*, 135–50.

Morse, P. M. and Feshbach, H. 1953. *Methods of theoretical physics*. McGraw-Hill, New York.

NASA Ames Research Center–Stanford University, Center for Turbulence Research. 1987. Studying turbulence using numerical simulation databases. Proceedings of the 1987 summer program, Report CTR-S87, December 1987.

1988. Studying turbulence using numerical simulation databases – II. Proceedings of the 1988 summer program, Report CTR-S88, December 1988.

Obremski, H. J., Morkovin, M., and Landahl, M. T. 1969. A portfolio of stability characteristics of incompressible boundary layers. AGARDOgraph 134, NATO, Paris.

Obukhov, A. M. 1941. Energy distribution in the spectrum of a turbulent flow. *Izvestya AN SSR, Ser. geogr. geofiz.*, No. 4–5, 453–66.

Orr, W. M. F. 1907. The stability or instability of the steady motions of a liquid. *Proc. R. Irish Acad. A 27*, 9–27, 69–138.

Orszag, S. A. and Patera, A. T. 1983. Secondary instability of wall-bounded shear flows. *J. Fluid Mech. 128*, 347–85.

Pao, Y. H. 1965. Structure of turbulent velocity and scalar fields at large wave numbers. *Phys. Fluids 8*, 1063–75.

Patel, V. C. 1965. Calibration of the Preston tube and limitations on its use in pressure gradients. *J. Fluid Mech. 23*, 185–208.

Patera, A., and Orzag, S. 1981. Subcritical transition to turbulence in plane shear flows. In *Transition to turbulence* (R. E. Meyer, ed.). Academic Press, New York, pp. 127–48.

Phillips, O. M. 1957. On the generation of waves by turbulent wind. *J. Fluid Mech. 4*, 426–34.

Pierrehumbert, R. T. and Widnall, S. E. 1982. The two- and three-dimensional instabilities of a spatially periodic shear layer. *J. Fluid Mech. 114*, 59–82.

Prandtl, L. 1925. Bericht über Untersuchingen zur ausgebildeten Turbulenz. *Z. Angew. Math. Mech. 5*, 136–7.

1945. Uber ein neues Formelsystem fur die ausgebildeten Turbulenz. *Nachr. Ges. Wiss. Göttingen, Math.-phys. Kl.*, 6–19.

Praturi, A. K. and Brodkey, R. S. 1978. A stereoscopic visual study of coherent structures in turbulent flow. *J. Fluid Mech. 89*, 251–72.

Ramamonjiarisoa, A. 1974. *Contribution a l'étude de la structure statistique et des mécanismes de génération des vagues de vent.* Thèse de doctorat d'état, Institut de Mécanique Statistique de la Turbulence, Université de Provence.

Rao, K. N., Narasimha, R., and Badri Narayanan, M. A. 1971. The "bursting" phenomenon in a turbulent boundary layer. *J. Fluid Mech. 48*, 339–52.

Rayleigh, Lord. 1878–79. On the instability of jets. *Proc. London Math. Soc. 10*, 4–13; *Scientific papers,* Vol. 1, 361–71. Cambridge University Press, Cambridge.

1880. On the stability, or instability, of certain fluid motions. *Proc. London Math. Soc. 11*, 57–70; *Scientific papers,* Vol. 1, pp. 474–87. Cambridge University Press, Cambridge.

Reynolds, O. 1883. An experimental investigation of the circumstances which determine whether the motion of water shall be direct or sinuous, and of the law of resistance in parallel channels. *Philos. Trans. R. Soc., London 174*, 935–82.

1894. On the dynamical theory of incompressible viscous fluids and the determination of the criterion. *Philos. Trans. R. Soc. London 186*, 123–61.

Ribner, H. S. 1957. Reflection, transmission, and amplification of sound by moving medium. *J. Acoust. Soc. Am. 29*, 435–41.

Robinson, S. K., Kline, S. J., and Spalart, P. R. 1988. Quasi-coherent structures in the turbulent boundary layer: Part II. Verification and new information from a numerically simulated flat-plate layer. In *Zoran P. Zaric' memorial international seminar on near-wall turbulence* (Dubrovnik, 1988).

Ross, J. A., Barnes, F. H., Burns, J. G., and Ross, M. A. S. 1970. The flat plate boundary layer. Part 3. Comparison of theory with experiment. *J. Fluid Mech. 43*, 819–32.

Ruelle, D. and Takens, F. 1971. On the nature of turbulence. *Commun. Math. Phys. 20*, 167–72.

Russell, J. M. and Landahl, M. T. 1984. The evolution of a flat eddy near a wall in an inviscid shear flow. *Phys. Fluids 27*, 557–70.

Saric, W. and Thomas, A. S. W. 1984. Experiments on the subharmonic route to turbulence in boundary layers. In *Turbulence and chaotic phenomena in fluids.* North-Holland, Amsterdam, pp. 117–22.

Schlichting, H. 1933. Zur Entstehung der Turbulenz bei der Plattenströmung. *Nachr. Ges. Wiss. Göttingen, Math.-phys. Kl.*, 181–208.

1935. Amplitudenverteilung und Energiebilanz der kleinen Störungen bei der Plattengrensschicht. *Nachr. Ges. Wiss. Göttingen, Math.-phys. Kl. 1*, 14–78.

Schubauer, G. 1954. Turbulent processes as observed in boundary layer and pipe. *J. Appl. Phys. 25*, 188–96.

Schubauer, G. B. and Klebanoff, P. S. 1951. Investigation of separation of the turbulent boundary layer. NACA, Rep. No. 1030.

Schubauer, G. B. and Skramstad, H. K. 1947. Laminar boundary layer oscillations and stability of laminar flow. *J. Aeronaut. Sci. 14*, 69–78.

Shen, S. F. 1954. Calculated amplified oscillations in plane Poiseuille flow and Blasius flows. *J. Aeronaut. Sci. 21*, 62–4.

Smagorinsky, J. S. 1963. General circulation experiments with the primitive equations. *Mon. Weather Rev. 91*, 99–165.

Smith, A. M. O. and Gamberoni, N. 1956. Transition, pressure gradient and stability theory. Douglas Aircraft Co. Rep. ES 26388.

Sommerfeld, A. 1908. Ein Beitrag zur hydrodynamischen Erklärung der turbulenten Flüssigkeitsbewegung. *Proceedings of the fourth international congress on mathematics* (Atti del IV congresso internazionale dei matematici, Roma, 1908), Vol. III, 116–124, Acc. dei Lincei, Rome 1909, 587 pp.

Spalart, P. R. 1988. Direct simulation of a turbulent boundary layer up to Re $\theta =$ 1410. *J. Fluid Mech. 187*, 61–98.

Sparrow, E. M., Husar, R. B., and Goldstein, R. J. 1970. Observations and other characteristics of thermals. *J. Fluid Mech. 41*, 793–800.

Squire, H. B. 1933. On the stability of the three-dimensional disturbances of viscous flow between parallel walls. *Proc. R. Soc. London, Ser. A 142*, 621–8.

Stokes, G. G. 1847. On the theory of oscillatory waves. *Trans. Camb. Philos. Soc. 8*, 441–73. also in ———. 1880. *Mathematical and physical papers*, Vol. I (J. Larmor, ed.). Cambridge University Press, Cambridge.

Su, M.-Y., Begin, M., Marler, P., and Myrick, R. 1982. Experiments on nonlinear instabilities and evolution of steep gravity wave trains. *J. Fluid Mech. 124*, 45–72.

Taylor, G. I. 1915. Eddy motion in the atmosphere. *Philos. Trans. R. Soc. London, Ser. A 215*, 1–26.

1923. Diffusion by continuous movements. *Proc. London Math. Soc. 20* (2), 196–211.

1935. Statistical theory of turbulence. Parts 1–4. *Proc. R. Soc. London, Ser. A 151*, 421–78.

1938. The spectrum of turbulence. *Proc. R. Soc. London, Ser. A 164*, 476–9.

Tennekes, H. and Lumley, J. L. 1972. *A first course in turbulence*. MIT Press, Cambridge, Mass.

Tietjens, O. 1925. Uber die Entstehung der Turbulenz. *Z. Angew. Math. Mech. 5*, 200–17.

Tollmien, W. 1929. Uber die Entstehung der Turbulenz. *Nachr. Ges. Wiss. Göttingen, Math.-phys. Kl.*, 21–44.

1935. Ein allgemeines Kriterium der Instabilität laminarer Geschwindigkeitsverteilungen. *Nachr. Ges. Wiss. Göttingen, Math.-phys. Kl. 50*, 79–114.

Townsend, A. A. 1956. *The structure of turbulent shear flow.* Cambridge University Press, Cambridge.

1976. *The structure of turbulent shear flow,* 2nd ed. Cambridge University Press, Cambridge.

Uberoi, M. S. 1954. Energy transfer in isotropic turbulence. *Phys. Fluids 6,* 1048–56.

Van Atta, C. W. and Chen, W. Y. 1969. Measurement of spectral energy transfer in grid turbulence. *J. Fluid Mech. 38,* 743–63.

van Ingen, J. L. 1956. A suggested semiempirical method for the calculation of the boundary layer transition region. Report VTH-74, Department of Aeronautical Engineering, University of Delft.

Von Karman, Th. and Howarth, L. 1938. On the statistical theory of turbulence. *Proc. R. Soc. London, Ser. A 164,* 192–215.

Whitham, G. B. 1965. A general approach to linear and non-linear dispersive waves using a Lagrangian. *J. Fluid Mech. 22,* 273–83.

1974. *Linear and nonlinear waves.* Wiley-Interscience, New York.

Widnall, S. E. 1984. Growth of turbulent spots in plane Poiseuille flow. In *Turbulence and chaotic phenomena in fluids* (T. Tatsumi, ed.). Elsevier, Amsterdam, pp. 93–8.

Willis, G. E. and Deardorff, J. W. 1967. Confirmation and renumbering of the discrete heat flux transitions of Malkus. *Phys. Fluids 10,* 1861–6.

Winant, C. D. and Browand, F. K. 1974. Vortex pairing: The mechanism of turbulent mixing-layer growth at moderate Reynolds number. *J. Fluid Mech. 63,* 237–55.

Wygnanski, I., Oster, D., Fiedler, H., and Dziomba, B. 1979. On the perseverance of a quasi-two-dimensional eddy-structure in a turbulent mixing layer. *J. Fluid Mech. 93,* 325–35.

Yuen, H. C. and Lake, B. M. 1975. Nonlinear deep water waves; theory and experiment. *Phys. Fluids 18,* 956–60.

1980. Instabilities of waves on deep water. *Ann. Rev. Fluid Mech. 12,* 303–34.

Zakharov, V. E. and Shabat, A. B. 1971. Exact theory of two-dimensional self-focusing and one-dimensional self-modulating waves in nonlinear media. *Zh. Eksp. Teor. Fiz. 61,* 118–34. (Translated in *Sov. Phys. JETP 34,* 62–7, 1972.)

Index

Page numbers followed by an "f" indicate figures.

167